Residential Code Essentials

Based on the 2018 International Residential Code®

International Code Council
Stephen A. Van Note

RESIDENTIAL CODE ESSENTIALS:
Based on the 2018 International
RESIDENTIAL Code®

Stephen A. Van Note

International Code Council

Executive Vice President and Director of Business Development:
Mark A. Johnson

Senior Vice President, Business and Product Development:
Hamid Naderi

Vice President and Technical Director, Products and Services:
Doug Thornburg

Senior Marketing Specialist:
Dianna Hallmark

ISBN: 978-1-60983-787-7

Cover Design:	Lisa Triska
Project Editor:	Greg Dickson
Project Head:	Steve Van Note
Publications Manager:	Mary Lou Luif
Production Technician:	Sue Brockman

First Printing: May 2018

PRINTED IN THE U.S.A.

Contents

PART VII : ENERGY CONSERVATION 241

PART VIII : PROTECTION FROM OTHER HAZARDS 253

x

Preface

Construction of residential buildings routinely consists of conventional practices, those tried-and-true methods that have performed well over the years and have long been recognized by the building code. With the introduction of new technology, materials, and methods, improved understanding of safe and healthy living environments, and innovation in dwelling designs, residential construction and the codes that regulate it have become increasingly complex. Such complexity is necessary to afford flexibility in design and construction. Reference publications intending to explain the provisions regulating residential construction may overwhelm the reader with a broad range of topics and alternatives, may provide superficial coverage of all requirements, or may focus on the details of a limited number of provisions.

Residential Code Essentials: Based on the 2018 International Residential Code® was developed to address the need for an illustrated text explaining the basics of the residential code—those provisions essential to understanding the application of the code to the most commonly encountered building practices. The text is presented and organized in a user-friendly manner with an emphasis on technical accuracy and clear non-code language. The content is directed to readers with a basic understanding of conventional dwelling construction but a less than complete knowledge of the *International Residential Code*® (IRC).

Anyone involved in the design, construction, or inspection of residential buildings will benefit from this book. Beginning and experienced inspectors, contractors, home builders, architects, designers, home inspectors and students of construction technology or related fields will gain a fundamental understanding and practical application of the frequently used provisions of the 2018 edition of the IRC.

The content of *Residential Code Essentials* is organized to correspond to the order of construction, beginning with sitework and foundations through completion of a safe, healthy and energy-efficient dwelling. Mechanical, fuel-gas, plumbing and electrical provisions are placed in separate chapters. The advantage of this format to the reader is that it pulls related information together from various sections of the IRC into one convenient location of the text and provides a familiar frame of reference to those with construction experience. The book explains the difference between "prescriptive" and "performance" requirements. Prescriptive structural design requirements to resist the forces of wind, earthquake and snow are described and illustrated in an easy-to-understand way. Structural topics include conventional footings and foundations (including the fundamentals of soil capacity), conventional wood floor, wall and roof framing, engineered wood products and seismic reinforcing of masonry chimneys. Fire- and life-safety concerns are addressed with topics including means of egress, emergency escape, stairways, fall protection, smoke alarms, fire sprinklers and fire-resistant construction. *Residential Code Essentials* also covers the minimum interior environmental conditions for a healthy living environment, weather protection and energy conservation measures.

Correct and reasonable application of the code provisions is enhanced by a basic understanding of the code development process, the scope, intent and correlation of the family of International Codes, and the proper administration of those codes. Such fundamental information is provided in the opening chapters of this publication. The book also explains the interaction of a building code with other local and state regulations and includes discussion of common hazards of the built environment that may be regulated by state or federal agencies.

This book does not intend to cover all provisions of the IRC or all of the accepted materials and methods of construction of residential buildings. Focusing in some detail on the most common conventional construction provisions affords an opportunity to fully understand the basics without exploring every variable and alternative. This is not to say that information not covered is any less important or valid. This book is best used as a companion to the IRC, which should be referenced for more complete information.

Residential Code Essentials features full-color illustrations to assist the reader in visualizing the application of the code requirements. Practical examples, simplified tables and highlights of particularly useful information also aid in understanding the provisions and determining code compliance. References to the applicable sections of the 2018 edition of the IRC are helpful in locating the corresponding code language and related topics in the code. A glossary of code and construction terms clarifies the meaning of the technical provisions.

ABOUT THE INTERNATIONAL RESIDENTIAL CODE

The IRC is a comprehensive, stand-alone residential code that establishes minimum regulations for the construction of one- and two-family dwellings and townhouses up to three stories in height, including provisions for fire and life safety, structural design, energy conservation and mechanical, fuel-gas, plumbing and electrical systems. The IRC incorporates prescriptive provisions for conventional construction as well as performance criteria that allow the use of new materials and new building designs.

The IRC is one of the codes in the family of International Codes published by the International Code Council (ICC). All are maintained and updated through an open code development process and are available internationally for adoption by the governing authority to provide consistent enforceable regulations for the built environment.

ACKNOWLEDGMENTS

Grateful appreciation is extended to Sandra Hyde, P.E., Senior Staff Engineer for ICC Product Development for her thorough review and valued comments in developing the 2018 edition of *Residential Code Essentials*. Previous editions (including *Building Code Basics, Residential*) were also the result of a collaborative effort. In addition to Sandra Hyde, valuable contributions were made by the following: Hamid Naderi, PE, Senior Vice President, ICC Business and Product Development, who developed the original concept and was instrumental in launching the series; Doug Thornburg, AIA, ICC Vice President and Technical Director of Products and Services, who provided expert direction and advice, as well as oversight in maintaining consistency throughout the series; John Henry, PE, former ICC Principal Staff Engineer, who offered his generous assistance and patient explanations relating to the structural provisions; and former ICC staff member Scott Stookey, who provided welcome expertise on fire resistance and fire protection systems. All contributed to the accuracy and quality of the finished product.

ABOUT THE AUTHOR

Stephen A. Van Note, CBO
International Code Council
Managing Director, Product Development

Steve Van Note is the Managing Director of Product Development for the International Code Council (ICC), where he is responsible for developing technical resource materials in support of the International Codes. His role also includes the management, review and technical editing of publications developed by ICC staff members and other expert authors. He has authored a number of ICC support publications, including *Significant Changes to the International Residential Code* and *Inspector Skills*. In addition, Steve develops and presents code seminars nationally. He has over 40 years of experience in the construction and building code arena. Prior to joining ICC in 2006, Steve was building official for Linn County, Iowa. Prior to his 15 years at Linn County, he was a carpenter and construction project manager for residential, commercial and industrial buildings. A certified building official and plans examiner, Steve also holds certifications in several inspection categories.

ABOUT THE INTERNATIONAL CODE COUNCIL

The International Code Council is a member-focused association. It is dedicated to developing model codes and standards used in the design, build and compliance process to construct safe, sustainable, affordable and resilient structures. Most U.S. communities and many global markets choose the International Codes. ICC Evaluation Service (ICC-ES) is the industry leader in performing technical evaluations for code compliance fostering safe and sustainable design and construction.

Headquarters:
500 New Jersey Avenue, NW, 6th Floor,
Washington, DC 20001-2070

Regional Offices:
Birmingham, AL; Chicago, IL; Los Angeles, CA

1-888-422-7233
www.iccsafe.org

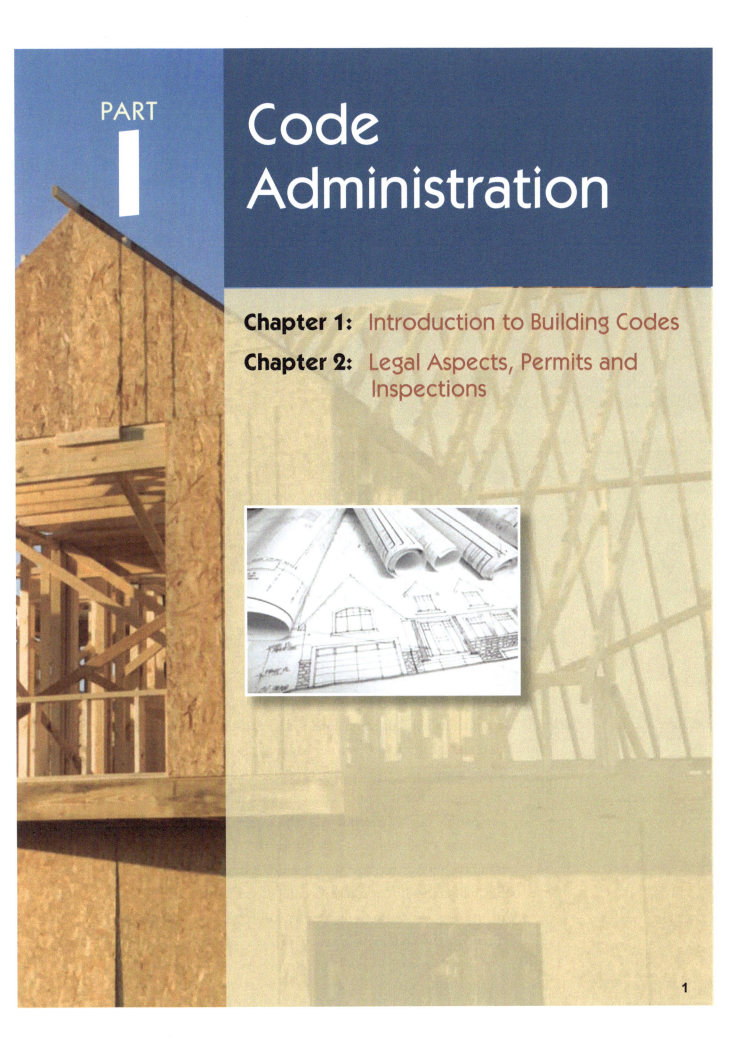

PART

1

Code Administration

Introduction to Building Codes

Building codes, in the broadest sense, are the various sets of regulations related to the construction, alteration, maintenance and use of buildings and structures. Sometimes collectively called construction codes, the separate volumes include not only structural considerations, but provisions for fire and life safety, energy conservation and systems for heating, cooling, plumbing and electrical utilities. These codes serve primarily to protect the safety and welfare of the building occupants and the public. One in a family of coordinated and compatible construction codes, the *International Residential Code* (IRC) combines all elements necessary for the construction of one- and two-family dwellings and townhouses into a single volume. Providing design flexibility, the IRC references companion International Codes for elements of construction outside the scope of the IRC. This chapter briefly discusses the code development process and the scope of some of the companion codes, followed by a more detailed examination of the IRC scope (Figure 1-1).

FIGURE 1-1 International Codes

CODE DEVELOPMENT

Construction technology, methods, materials, equipment and processes are constantly evolving. Accordingly, the International Code Council model codes (I-Codes) are systematically revised at periodic intervals to keep up with technical changes and to address the improved understanding of hazards. The I-Codes are revised and updated through an open process that invites participation by all stakeholders and affected parties. This code development process can involve exhaustive research, review, discussion and debate of the potential changes.

A code change begins with the submittal of a proposal to add, revise or delete a code provision. Any interested individual or group (other than ICC staff) may submit a code change proposal and participate in the proceedings in which it and all other proposals are considered. Proposed code changes are submitted online through the ICC cdpACCESS™ website (code development process access). Following the publication and distribution of the proposals, an open public hearing, the Committee Action Hearing (CAH), is held before a committee comprising representatives from across the construction industry and government, including code officials, contractors, builders, architects, engineers, industry professionals and others with expertise related to the applicable code or portion of the code under consideration. This open debate and broad participation before the committee reflect a consensus of the construction community and those impacted by building and fire codes in the decision-making process. The committee may approve, modify or disapprove the code change proposal.

If there is a successful motion from the floor regarding the committee's action, then the assembly floor motion will become available for online balloting after the Committee Action Hearing concludes. All members of ICC can participate in the Online Assembly Floor-Motion Votes. If the assembly floor motion receives a majority vote, then the motion becomes a Public Comment to be heard at the Public Comment Hearing. Once the first stage of online voting is completed, the results of the Committee Action Hearing and the subsequent online voting are compiled and posted as the Report of Committee Action Hearing (ROCAH). Following the availability of the ROCAH, anyone may submit a public comment proposing to modify or overturn the hearing results.

The next public hearing is the Public Comment Hearing, in which the merits of code change proposals that received public comment are debated. Though any interested party may offer testimony, only ICC Governmental Member Voting Representatives (designated public safety officials of a government jurisdiction responsible for administering and enforcing codes) and ICC Honorary Members are permitted to cast votes at the Public Comment Hearing. Public safety officials have no vested financial interest in the outcome, and they legitimately represent the public interest. Votes are cast electronically at the Public Comment Hearing, and the results are then made available online at cdpACCESS. In the Online Governmental Consensus Vote, again only the ICC Governmental Member Voting Representatives and ICC Honorary Members are permitted to vote. The Online Governmental Consensus Vote determines the revisions to the next edition of the code.

A new edition of the code is published every three years. Each I-Code is developed during a single twelve-month cycle during the three-year period. There are two separate code change cycles within each three-year period, with individual codes assigned to one of the two cycles. This important process ensures that the International Codes reflect the latest technical advances and address the concerns of those throughout the industry in a fair and equitable manner.

THE BUILDING CODES: SCOPE AND LIMITATIONS

A number of features are common to all of the International Codes. Each code begins by stating its scope of application. The scope establishes the range of buildings, uses, construction, equipment, and systems to which the particular code applies. The International Building, Mechanical, Fuel Gas and Plumbing codes exclude from their scope those residential buildings regulated by the IRC. A purpose statement follows the scope and includes the intent to provide minimum standards to protect the health, safety and welfare of the public and the occupant or user of the space or building. Subsequent sections place limitations on the application of the code. For example, each code permits the continued legal use of existing buildings. A building does not need to be brought into compliance with the current codes provided the building does not create hazards to the occupants or property and meets minimum acceptable standards for health and sanitation. In the case of an addition to an existing building, only the addition need comply with the current code, provided it does not cause an unsafe condition in the existing structure. Each International Code also references other codes and standards for use under specific circumstances. For example, the IRC references the *International Building Code* (IBC) for an engineered design of structural elements beyond the scope of the IRC. Finally, the appendices of each code are not in effect unless they are specifically adopted by the authority having jurisdiction (see Chapter 2).

International Building Code® (IBC®)

The provisions of the IBC apply to the construction, alteration, maintenance, use and occupancy of all buildings and structures except detached one- and two-family dwellings and townhouses and their accessory structures, which are covered by the IRC. In addition to structural components and systems, the IBC provides for a safe means of egress, accessibility for persons with disabilities, fire resistance, fire protection systems, weather resistance, finishes and interior environments. These regulations are typically related to the use and occupancy of the building. That is, the IBC assesses relative risks or hazards based on the functions within the building and controls design accordingly. As a result, provisions regulating the building's size, means of egress, fire-resistive elements and fire protection systems vary significantly among sports arenas, hospitals, schools, apartments and office buildings (Figure 1-3).

FIGURE 1-2 Apartment building constructed under the provisions of the IBC

International Mechanical Code® (IMC®)

The provisions of the IMC generally apply to the installation, alteration, use and maintenance of permanent mechanical systems utilized for comfort heating, cooling and ventilation, and other mechanical processes within buildings.

International Fuel Gas Code® (IFGC®)

The IFGC regulates the installation of natural gas and liquefied petroleum (LP)-gas piping systems, fuel gas appliances, gaseous hydrogen systems and related accessories. Coverage includes pipe sizing and other design considerations, approved materials, installation, testing, inspection, operation and maintenance. The equipment installation requirements include combustion and ventilation air, approved venting and connection to the fuel gas piping system.

International Plumbing Code® (IPC®)

The provisions of the IPC generally apply to the installation, alteration, use, and maintenance of plumbing systems. The IPC includes the material and installation requirements for water supply and distribution, plumbing fixtures, drain-waste and vent (DWV) piping and storm drainage systems.

International Fire Code® (IFC®)

The IFC establishes regulations for the prevention of fire and explosion hazards arising from the storage and use of hazardous materials, and it controls other conditions that are hazardous to life, property or public welfare. It also provides for the safety of fire fighters during emergency operations. Storage, building use and building maintenance provisions apply during the life of buildings. The IFC also contains design and construction provisions related to fire department access, fire protection water supply, and fire alarm and fire protection systems.

International Property Maintenance Code® (IPMC®)

The IPMC provides for the regulation and safe use of existing structures in the interest of the social and economic welfare of the community. It establishes minimum maintenance standards for light, ventilation, space, heating, sanitation, protection from the elements, life safety and safety from fires and other hazards.

International Existing Building Code® (IEBC®)

The IEBC intends to provide flexibility and alternative approaches in the repair, alteration, addition to, relocation and change of use of existing buildings and still safeguard the public health, safety and welfare. The alternatives offered to the applicant include prescriptive, work-area and performance-compliance methods.

International Energy Conservation Code® (IECC®)

The IECC regulates the design and construction of residential and commercial buildings for the effective use and conservation of energy. The IECC intends to provide flexibility in the methods of compliance with energy conservation requirements.

International Swimming Pool and Spa Code® (ISPSC®)

The ISPSC is a comprehensive swimming pool code coordinated with the current requirements in the International Codes and the Association of Pool & Spa Professionals (APSP) standards. The ISPSC establishes minimum regulations for public and residential pools, spas and hot tubs using prescriptive and performance-related provisions.

INTERNATIONAL RESIDENTIAL CODE® (IRC®)

The provisions of the IRC generally apply to the construction, alteration, use and occupancy of detached one- and two-family dwellings and townhouses (Figures 1-3, 1-4 and 1-5). Such buildings are limited to not more than three stories above grade plane in height,

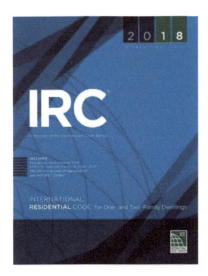

FIGURE 1-3 2018 *International Residential Code* (IRC)

and each dwelling unit must have a separate means of egress. This comprehensive, stand-alone residential code includes provisions for structural elements; fire and life safety; a healthy living environment; energy conservation; and mechanical, fuel gas, plumbing and electrical systems. The IRC incorporates prescriptive provisions for conventional light frame construction as well as performance criteria that allow the use of new materials and new building designs (see Chapter 4). As in the other International Codes, the purpose of the IRC is to safeguard the public safety, health and general welfare from fire and other potential hazards attributed to the built environment. The code establishes minimum requirements, reasonably balanced for affordability, that provide strong, stable and sanitary homes that conserve energy while still offering adequate lighting, comfort conditioning, and ventilation. **[Ref. R101, R202]**

Dwellings and townhouses

The building height and means of egress requirements of the IRC apply equally to one- and two-family dwellings and townhouses. Although the code generally limits these residential buildings to three stories above ground level, this still permits a full basement in addition to three stories above, effectively creating a building with four floor levels. In addition, the IRC permits a habitable attic, which is not counted as a story, conceivably creating a fifth habitable level, though such an installation is not common (Figure 1-6). As will be seen in later chapters, structural and other design criteria of the code may further limit the height and number of stories of the building. The code does not limit the total area of dwellings, however.

In addition to height considerations, each dwelling unit requires its own separate means of exiting the building to the outdoors (see Chapter 8). Only one exterior exit door is

FIGURE 1-4 Detached single-family dwelling regulated by the IRC

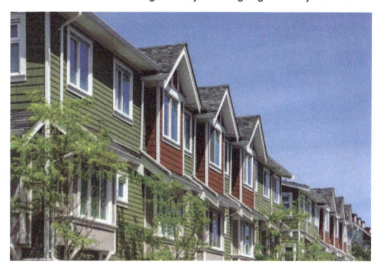
FIGURE 1-5 Townhouses

FIGURE 1-6 Three stories above grade plane with basement and habitable attic

Code Essentials

IRC dwelling unit limits
- Three stories above grade plane
- Separate means of egress

No limits on
- Number of townhouses
- Dwelling unit area
- Travel distance to exit ●

required, and the travel distance to that exit is not regulated, no matter the size or number of stories of the dwelling unit. Two-family dwellings (Figure 1-7) and townhouses require fire-resistant separations between dwelling units. Limited protection against the spread of fire is also required between a dwelling unit and an attached garage (see Chapter 9). **[Ref. R101.2, R202]**

FIGURE 1-7 Detached two-family dwelling regulated by the IRC

The IRC does not limit the number of townhouses in a group of townhouses but does require the building to satisfy certain other conditions. To qualify as a townhouse, there must be at least three attached dwelling units, and each unit must run from foundation to roof. That is, any portion of a townhouse is not permitted to be placed above any portion of another townhouse. In addition, each townhouse must be open to a yard or public way on at least two sides (Figure 1-8). Multifamily dwellings that do not meet the definition of townhouses fall under the provisions of the IBC (Figure 1-9). **[Ref. R202]**

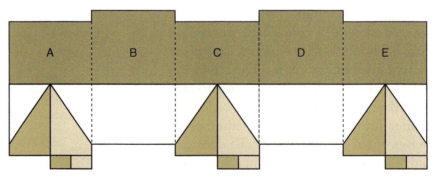

FIGURE 1-8 Townhouses open on front and back

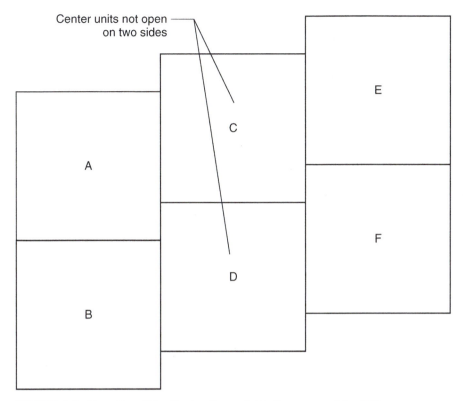

FIGURE 1-9 Six-unit multifamily dwelling outside the scope of the IRC

Manufactured homes

Regulations related to manufactured homes appear in Appendix E of the IRC and are not in effect unless specifically adopted by the jurisdiction. For purposes of the IRC, a manufactured home is considered the same as a mobile home, though the preferred term since 1974 federal legislation is *manufactured home*. The United States Department of Housing and Urban Development (HUD) regulates the construction of manufactured homes, which are built in a manufacturing plant to comply with the *Manufactured Home Construction and Safety Standards* (HUD code). Compliance is verified through state or third-party inspection agencies. Each transportable section must display a red certification label on the exterior.

Manufactured homes are built on a permanent chassis and are designed to be used as a dwelling with or without a permanent foundation. This design assures transportability for relocation and differentiates manufactured homes from modular homes and other factory-built, panelized, or component structures.

Local jurisdictions have no authority to regulate the design or construction of a HUD-regulated manufactured home. IRC Appendix E applies to on-site construction related to manufactured homes used as single-family dwellings and installed on privately owned lots. The provisions relate to the foundation system, connection to utility services (water, sewer, fuel and electricity), and alterations, additions or repairs to the manufactured home. However, any modification or addition must be in compliance with HUD regulations and cannot proceed if otherwise prohibited. Attachment of an accessory building, such as a garage, is generally prohibited unless the design is substantiated through engineering calculations. Factory-built structures that do not fall under the HUD rules are regulated by local jurisdictions like any other building, or in some cases are regulated by a state agency.

Accessory buildings and structures

Accessory buildings must be located on the same lot as the dwelling and be considered accessory to and incidental to the dwelling. Detached garages and sheds are the most common examples of accessory buildings (Figure 1-10). Workshops, storage buildings, gazebos, playhouses, swimming pool equipment rooms and garden buildings are other examples of structures typically considered accessory to dwellings. **[Ref. R101.2, R202]**

> ### Code Essentials
>
> **Accessory buildings**
>
> - Maximum 3 stories
> - Unlimited in area
> - Use incidental and accessory to dwelling
> - On same lot as dwelling ●

FIGURE 1-10 Accessory building

Existing structures

Provisions allowing the legal occupancy of buildings to continue without fully complying with current codes are often referred to as grandfathering or grandfather clauses. As with other International Codes, the IRC provides such relief for existing buildings. To impose regulations to bring existing buildings into current compliance would be impractical and unreasonable and would penalize owners of buildings that complied with applicable laws at the time of their construction. Of course, if, due to lack of repair or maintenance, buildings fall below the generally acceptable threshold for sanitation, health, safety and welfare of the occupants and the public, the IRC requires corrections in accordance with specific code provisions and the referenced provisions of the IPMC or the IFC.

These grandfathering provisions not only apply to the continued use and occupancy of an existing building in the absence of construction activity, they also apply to existing buildings undergoing modifications or additions. Generally, only the modification or addition need comply with the current code. There are some exceptions to this rule. For example, there are retroactive provisions for smoke alarms and carbon monoxide alarms when work requiring a permit occurs in existing buildings. (See Chapters 9 and 10.) The installation of smoke alarms, of paramount importance in saving lives, is considered reasonable and practical during construction work on existing buildings. Additions, alterations or repairs cannot cause any portion of the existing building to become unsafe or otherwise adversely affect the performance of the building. If an addition impeded the means of egress to the outdoors, added excessive loading to existing structural members, overloaded the electrical service, or exceeded the capacity of the plumbing drain, waste and vent (DWV) system, then any of the affected elements would need to be brought into compliance with the current code.

IRC Appendix J, if specifically adopted, does offer alternatives for compliance with the code during renovation of existing buildings. These provisions, similar to those found in the IEBC, intend to encourage the continued use or reuse of legally existing buildings and structures. Under Appendix J, construction work is categorized as repair, renovation, alteration or reconstruction. As the extent of work becomes greater, resulting in a higher category, the requirements become more stringent. While maintaining an acceptable level of safety, the alternative approaches offer a number of benefits to the owner and builder, such as allowing smaller dimensions for existing windows, doors, stair headroom and ceiling heights without modification. In some cases, certain nonconforming elements in sound condition, such as stairs and railings, may remain without modification even within prescribed work areas. [Ref. R102.7]

Legal Aspects, Permits and Inspections

Building codes intend to protect the health, safety and welfare of the public by establishing minimum acceptable standards of construction. In order to be effective, such codes must be legally adopted by the governmental jurisdiction and enforced by qualified officials appointed by the governing authority. This chapter explores the process of adopting, amending, and administering the *International Residential Code* (IRC), but the principles generally apply to the other International Codes as well.

CODE ADOPTION

The IRC and the family of International Codes are referred to as *model codes,* nationally recognized construction regulations that serve as models for local ordinances and are typically adopted by reference. The International Codes are maintained and updated through an open process of code development. Such a process relies on the participation of experts and all interested parties in the various fields of design, construction, manufacturing, government and code administration. These model codes are updated on three-year cycles to recognize new and developing technology, materials, and methods of construction. Changes in the codes are also often in response to natural or human-made disasters involving the loss of lives or the destruction of property.

Adoption of the IRC

The IRC becomes an enforceable regulation through the legal proceedings of the governmental jurisdiction. The adopting ordinance references the edition and title of the IRC and provides other necessary factual information (Figure 2-1). Typically the legislation also includes the purpose, scope, and effective date for the ordinance. As part of the adoption process, the government authority must also provide local information for insertion into code text, including the name of the jurisdiction, design criteria (see Chapter 4) and building sewer depths (see Chapter 13). **[Ref. R101.1, Table R301.2(1), P2603.5.1]**

ORDINANCE NO. 000-18
Residential Building Regulations

An ordinance of the city adopting the 2018 edition of the International Residential Code, regulating and governing the construction, alteration, movement, enlargement, replacement, repair, equipment, location, removal and demolition of detached one and two family dwellings and townhouses not more than three stories in height with separate means of egress and their accessory structures in the city providing for the issuance of permits and collection of fees therefore; repealing Ordinance No. 000-18 of the city and all other ordinances and parts of the ordinances in conflict therewith.

The city council does ordain as follows:

I. Except as hereafter modified, that the International Residential Code, 2018 Edition, including Appendix chapters _____, as published by the International Code Council, is hereby specifically adopted by reference
as the residential code of the city. The provisions of said residential code shall be controlling in the design, construction, quality of materials, erection, installation, addition, alteration, repair, location, relocation, replacement, removal, demolition, use and maintenance of detached one and two family dwellings and townhouses not more than three stories in height with separate means of egress and their accessory structures in the city and in all matters covered by said residential code within the city.

II. The following sections are hereby revised:
Section R101.1. Insert: [NAME OF JURISDICTION]
Table R301.2(1) Insert: [APPROPRIATE DESIGN CRITERIA]
Section P2603.5.1 Insert: [NUMBER OF INCHES IN TWO LOCATIONS]

III. That this ordinance and the rules, regulations, provisions, requirements, orders and matters established and adopted hereby shall take effect and be in full force and effect upon the date of its final passage and adoption.

FIGURE 2-1 Sample ordinance adopting the IRC

Amending the IRC

Advantages to the adoption of model codes include consistency, uniformity, and correlation among construction codes and across jurisdictional boundaries. Such uniformity benefits designers and builders as well as government officials in charge of administering the codes. Although states, cities, counties, and similar governmental jurisdictions typically do not have the resources to develop and maintain comprehensive construction codes, jurisdictions do have the ability to modify the model code through amendments placed in the adopting ordinance. Excessive local amendments to adopted model codes may be contrary to the goals of consistency and may offset the advantages and legal defensibility of nationally recognized standards. Still, some modification of the IRC may be necessary or desirable for any number of reasons.

Though model codes anticipate location and climate differences, amendments to them may be influenced by unique characteristics and conditions, such as geographic location, weather, topography, flooding, soil properties and water tables. There also may be considerations of political influences, local traditions or customs, compatibility issues with other state or local laws, or the existence of unique housing stock, such as in historic districts. It is important to recognize that these amendments must be legally instituted with care through the adopting ordinance and laws of the jurisdiction. Otherwise, actions to enforce provisions not contained in the IRC or to waive certain requirements that are in the model code, in addition to creating inconsistency in the application of the code, will not be in accordance with the law.

Appendices

The appendices are developed in much the same manner as the main body of the model code. However, the appendix information is judged outside the scope and purpose of the code at the time of code publication. Many times an appendix offers supplemental information, alternative methods, or recommended procedures. The information may also be specialized and applicable or of interest to only a limited number of jurisdictions. Although an appendix may provide some guidelines or examples of recommended practices or assist in the determination of alternative materials or methods, it will have no legal status and cannot be enforced until it is specifically recognized in the adopting legislation. Appendix chapters or portions of such chapters that gain general acceptance over time are sometimes moved into the main body of the model code through the code development process. For example, provisions for gray water recycling systems that were once located in Appendix O are now part of the nonpotable water provisions of Chapter 29 of the IRC. [Ref. R102.5]

Local and state laws

The IRC is not meant to nullify any local, state or federal law, and in many cases, such other laws supersede provisions found in the model code. Careful consideration should be given to the interaction of the residential code with other codes and laws of the jurisdiction and the state during the process of adoption. Zoning ordinances, for example, may include more restrictive provisions on distances to property lines and between structures, permissible heights and areas of buildings, the number of buildings on a lot, and the specific conditions of use and occupancy of residences and accessory buildings. Other local ordinances impacting the residential code and construction may include those regulating public streets, storm water management, erosion control, public and private sewers and private wells. State laws may override the building codes in matters related to energy conservation, manufactured and modular housing, and accessibility for persons with disabilities. State law also often determines the circumstances under which a licensed engineer or architect is required and sets the licensing regulations for these design professionals (Figure 2-2). [Ref. R102.2]

FIGURE 2-2 Engineer's seal

AUTHORITY

The IRC establishes a department of building safety, commonly referred to as the building department, and designates the officer in charge of the administration and enforcement of the code as the building official. The appointing authority of the jurisdiction appoints the building official, though jurisdiction job titles may vary. The building official in turn assigns some degree of decision-making authority to deputies, plans examiners, inspectors, permit technicians and other employees. The role of a building official in protecting public safety is complex and challenging. It follows that the position demands skills, knowledge, and abilities to not only fulfill the duties, but elevate the credibility of the department in the eyes of the public.

Authority and duties of the building official

The IRC charges the building official with enforcing the provisions of the code and assigns broad authority and discretion to do so. With discretion comes the responsibility to make decisions in keeping with the intent of the IRC. The building official does not have authority to waive code requirements. In the same way, the building official has no authority to require more than the code stipulates. The IRC also authorizes the building official to develop policies and procedures for consistent application of the code provisions. [Ref. R104.1]

Code Essentials

Building official duties

- Enforce the code
- Review construction documents
- Issue permits
- Issue notices and orders
- Conduct inspections
- Maintain records

Building official authority

- Make interpretations
- Adopt policies and procedures
- Approve modifications and alternatives

Limits on authority

- Cannot waive code requirements
- Cannot require more than the code ●

In effectively performing the duties listed in the code, the building official must also have an understanding of the legal aspects of code administration. While given broad authority for enforcement, including the issuance of notices and orders, the building official must also recognize the rights of due process afforded to the public. Equally important to the building department in securing safe buildings for the community is to build the public trust through communication, respect, and fairness, so that the department is viewed as a resource rather than an adversary.

Interpretations

Many code provisions are clear and easily understood. The dimensions for stair rise and run, handrail height, and spacing for openings in guards, for example, are specific and objectively measurable. Such provisions are referred to as *prescriptive*—a clear set of rules to follow to gain compliance. The shape of other than round handrails, on the other hand, is less clearly defined. Although the code sets some parameters for dimensions of these handrails, it also permits any handrail that provides equivalent graspability.

The term "equivalent graspability" is a performance requirement, meaning that an element must function to satisfy certain acceptable criteria. Determination of compliance requires some level of judgment on the part of the building official. Though the IRC intends to be largely prescriptive in nature, it purposely offers performance criteria as well, to allow flexibility in design and construction and to not favor certain materials or methods over any other.

Alternative methods and materials and Evaluation Service reports

The IRC is specific in its intention to not exclude the use of any material or method of construction, even if such methods are not specifically described by the code, subject to approval by the building official. The building official has not only the authority but an obligation to approve such alternatives where he or she finds that the proposed material or construction meets the intent of the IRC and is equivalent to the code provisions. When a proposal for using an alternative method or material is denied, the IRC requires the building official to state the reason for disapproval in writing. With modern technology advancing at a record pace, new and innovative building products are continuously introduced to the market on a global scale. Reports issued by the International Code Council's Evaluation Service (ICC-ES) are valuable resources in verifying performance equal to the code requirements (Figure 2-3). In the absence of ICC-ES evaluation reports and where insufficient data or documentation exists, the building official may require that tests be performed by an approved agency to demonstrate compliance with the code. Also, compliance with the specific performance-based provisions of the referenced International Codes satisfies the IRC requirements. The performance and alternative-methods provisions of the IRC and the use of the ICC Evaluation Reports provide for flexibility and encourage innovative and new materials, design and construction while protecting the public safety. All published ICC Evaluation Reports are available free of charge and can be accessed online at www.icc-es.org. **[Ref. R104.11]**

Occasionally there are instances where it is not feasible to fully comply with the strict letter of the code. In this case, the IRC allows a modification of the code provision for individual cases when approved by the building official. The decision is based on docu-

You Should Know

Alternative methods and materials

- The building official approves alternatives that comply with the intent of the code.

ICC Evaluation Service (ES) Reports

- ES Reports are valuable tools for verifying that alternative methods and materials perform satisfactorily and are equivalent to those prescribed by the code. ●

ICC-ES Evaluation Report   **ESR-5000**

Issued January 2018
This report is subject to renewal January 2019.

www.icc-es.org | (800) 423-6587 | (562) 699-0543   *A Subsidiary of the International Code Council®*

DIVISION: 07 00 00—THERMAL AND MOISTURE PROTECTION
Section: 07 30 05—Roofing Felt and Underlayment

REPORT HOLDER:

ACME UNDERLAYMENTS UNLIMITED
52380 FLOWER STREET
CHICO, MONTANA 43820
(808) 664-1512
www.underlaymentunlimited.com

EVALUATION SUBJECT:

UU 100 UNDERLAYMENT FOR ASPHALT SHINGLE ROOF COVERINGS IN SEVERE CLIMATE AREAS

1.0 EVALUATION SCOPE

Compliance with the following codes:

- 2018 and 2015 *International Building Code®* (IBC)
- 2018 and 2015 *International Residential Code®* (IRC)

Property evaluated:

Ice barrier

2.0 USES

UU 100 is a self-adhering, rubberized asphalt membrane, complying with ASTM D1970, that is used over plywood substrates as ice barriers as specified in Chapter 15 of the IBC and Chapter 9 of the IRC.

3.0 DESCRIPTION

The UU 100 has a granule surfacing. The membrane has a silicone-treated release paper on the back that is removed prior to attachment to plywood sheathing. The membrane is a minimum of 0.040 inch (1.02 mm) thick and is supplied in rolls 36 inches (914 mm) wide and 66.7 feet (20.3 m) long.

4.0 INSTALLATION

Installation of the UU 100 membrane must comply with this report and the manufacturer's published installation instructions. The manufacturer's published installation instructions and this report must be available at the jobsite at all times during the installation.

Prior to application of the membrane, the deck surface must be free of frost, dust and dirt, loose nails and other protrusions. Damaged sheathing must be replaced. Installation is limited to plywood substrates. The membrane is designed for applications when the ambient air temperature is above 40°F (4.4°C).

Vertical ends and horizontal edges must be overlapped a minimum, respectively, of 6 inches (152 mm) and 3½ inches (89 mm). Horizontal edge overlaps must run with the flow of water in a shingling effect. A minimum of two layers of the membrane must be applied, starting at the lower edge (eave) of the roof, and extend a minimum of 24 inches (610 mm) inside the exterior wall line of the building. Final coverage width must comply with the code.

Installation of the roof covering can proceed immediately following application of the membrane. The membrane must be covered by an approved roof covering as soon as possible. For reroofing applications, the same procedures apply after removal of the old roof covering and roofing felts to expose the plywood roof deck.

5.0 CONDITIONS OF USE

The UU 100 membrane described in this report comply with, or are suitable alternatives to what is specified in, those codes listed in Section 1.0 of this report, subject to the following conditions:

5.1 Installation must comply with this report and the manufacturer's published installation instructions. In the event of a conflict between the manufacturer's published installation instructions and this report, this report governs.

5.2 Installation is limited to use on plywood substrates on structures located in areas where non-classified roof coverings are permitted.

5.3 Installation is limited to use with roof coverings that are mechanically fastened through the underlayment to the sheathing or rafters.

5.4 Installation is limited to roofs with ventilated attic spaces, in accordance with the requirements of the applicable code.

6.0 EVIDENCE SUBMITTED

6.1 Data in accordance with the ICC-ES Acceptance Criteria for Self-adhered Roof Underlayment for Use as an Ice Barrier in Severe Climate Areas (AC48), dated February 2012 (editorially revised December 2015).

6.2 Reports of testing in accordance with ASTM D1970.

7.0 IDENTIFICATION

The membrane is identified by labels on the rolls or packaging, displaying the Acme Underlayments Unlimited's name and address, the product name, and the evaluation report number ESR-5000.

FIGURE 2-3 Sample ICC-ES Evaluation Report

mentation demonstrating that the modification complies with the intent and purpose of the code and does not lessen health, safety or structural requirements. The building official records the decision in the building department files. **[Ref. R104.10]**

PERMITS

Except for a short list of work of a minor nature, any construction requires a permit before work begins, including work for the relocation or demolition of buildings. Work exempt from permits must still comply with the applicable IRC requirements. **[Ref. R105]**

Permit application

The owner or authorized agent makes application for a permit on a form furnished by the building department. In addition to providing a legal description of the property, the permit application must include the description of the work, the valuation of the proposed work, the use of the building and the applicant's signature (Figure 2-4). **[Ref. R105.1, R105.3]**

FIGURE 2-4 Sample application for permit form

Plans and specifications

Construction drawings and other submittal documents must accompany the permit application and be of sufficient detail and clarity to verify compliance with the code. The code also requires a site plan showing all new and existing structures with distances to lot lines. The extent of construction documents varies with the complexity and scope of the project. The building official is authorized to waive submittal documents for work of a minor nature, provided that code compliance can be verified by other means. Local or state laws determine requirements for a registered design professional to prepare the construction documents. In the case of any special conditions as determined by the building official, such as the sizing of a steel beam or the support for a concentrated load, the building official is also authorized to require plans to be prepared by a registered architect or engineer. [Ref. R106]

Residential Plan Review Checklist
2018 International Residential Code

	Foundations & Concrete—IRC Chapter 4	Section
☐	Lots graded to drain surface water away from foundation walls ≥ 6 inches fall within the first 10 feet	R401.3
☐	Concrete minimum specified compressive strength: • Footings, interior slabs: 2500 psi • Walls exposed to weather: 3000 psi • Garage slabs and exterior slabs 3500 air-entrained	Table R402.2
☐	Footings supported on undisturbed natural soils at least 12 inches below undisturbed ground or on engineered fill. Footing sizes: • Spread • Trench • Mat, pier, post, fireplace	R401.2, R403.1
☐	Foundation walls extend above the finished grade ≥ 6 inches (4 inches where masonry veneer is used)	R404.1.6
☐	Foundation anchor bolts ≥ ½ in. diameter extending ≥ 7 in. into masonry or concrete, maximum 6 ft OC and within 12" of ends	R403.1.6
☐	Concrete slab-on-ground floors ≥ 3.5 in. thick	R506.1
☐	Approved vapor retarder under slab	R506.2.3
☐	Approved drainage pipe at or below the area to be protected on a ≥ 2 inches of ¾ inch minimum washed crushed rock and covered with ≥ 6 inches of the same material.	R405.1
☐	Basement walls dampproofed (waterproofed if high water table)	R406.1, R406.2
	Floors—IRC Chapter 5	
☐	Spans for wood floor joists	Tables R502.3.1 (1) & (2)
☐	Spans of girders	Tables R602.7 (1) & (2)
☐	End bearing of joist beam or girder: • ≥ 1.5 inches on wood or metal • ≥ 3 inches on masonry or concrete • Or supported by approved joist hangers	R502.6
☐	Joists framing from opposite sides over bearing support shall lap ≥ 3 in.	R502.6.1
☐	Manufactured floor I-joist shall be installed in accordance with manufacturer's instructions.	
☐	Engineered floor truss design drawings and location drawing: • Hanger type & location • Approved connections • Bracing per engineered truss design drawings • Trusses shall not be cut, notched, spliced or altered	R502.11
☐	Draftstops when usable space above and below the concealed space of a floor/ceiling assembly with area ≤ 1000 square feet and approximately equal areas	R502.12, R302.12

FIGURE 2-5 Sample portion of a plan review checklist

A detailed review of plans and specifications is necessary to verify that the design complies with the code, thereby avoiding costly modifications during the course of construction. Jurisdictional policies differ in the handling of incomplete or incorrect plans. Depending on the complexity of the project and the significance of errors or omissions, the plans examiner, on behalf of the building official, may furnish comments or a list of code requirements to the applicant, may request supplemental information from the applicant, or may reject the plans and require submittal of revised documents (Figures 2-5 and 2-6).

FIGURE 2-6 Approved plans stamp

Fees

The jurisdiction establishes a schedule of fees at a level sufficient to offset the costs of providing the associated services to the public, including administration, plan review and inspection. Permit fees are often based on the total value of the work included in the scope of the permit, though other methods may be used. In any case, the building official should develop an equitable and consistent procedure for assessing fees related to the permit process (Figure 2-7). **[Ref. R108]**

City Building Department
PERMIT FEE SCHEDULE
As adopted by city resolution number 00-0000 effective ___/___/___

Building Department
City Of

TOTAL VALUATION	PERMIT FEE
$1 to $500	$24.00
$501 to $2,000	$24.00 for the first $500; plus $3 for each additional $100 or fraction thereof, to and including $2,000
$2,001 to $40,000	$69.00 for the first $2,000; plus $11 for each additional $1,000 or fraction thereof, to and including $40,000
$40,001 to $100,000	$487 for the first $40,000; plus $9 for each additional $1,000 or fraction thereof, to and including $100,000
$100,001 to $500,000	$1,027 for the first $100,000; plus $7 for each additional $1,000 or fraction thereof, to and including $500,000
$500,001 to $1,000,000	$3,827 for the first $500,000; plus $5 for each additional $1,000 or fraction thereof, to and including $1,000,000
$1,000,001 to $5,000,000	$6,327 for the first $1,000,000; plus $3 for each additional $1,000 or fraction thereof, to and including $5,000,000
$5,000,001 and up	$18,327 for the first $ 5,000,000; plus $1 for each additional $1,000 or fraction thereof

The applicant shall state the valuation of the proposed work at the time of application for a permit. The building official shall make the final determination of valuation in accordance with established department guidelines and published cost data such as ICC Building Valuation Data. Building permit valuation shall include total value of the work for which a permit is being issued, such as electrical, gas, mechanical, plumbing equipment and other permanent systems, including materials and labor.

Refunds shall be in accordance with the refund policy established by the building official in accordance with IRC Section R108.5.

FIGURE 2-7 Sample permit fee schedule

Permit issuance

In the language of the code, the building official must review the application and construction documents within a reasonable time and, when approved, issue the permit as soon as is practicable (Figure 2-8). The length of time considered reasonable will vary based on several factors, such as the complexity of the project and the completeness of the construction documents. The intent of the code is to allow sufficient time for a thorough review of plans to determine code compliance and to issue the permit in a timely manner to avoid unnecessary delays in the construction schedule, resulting in hardship to the applicant. A copy of the permit and the approved construction documents must be kept on the jobsite until completion of the project. [Ref. R105.3.1, R105.7, R106.3.1]

> **Code Essentials**
>
> **Permit-holder responsibilities**
>
> - Permit on jobsite
> - Approved plans on jobsite
> - Call for inspection
> - Provide access for inspection

Building Department
City Of

Notice
Building Permit

Permit No. 00-0000

Has been issued to

Owner ______________________

Contractor ______________________

For (Work Description) ______________________

At (Project Address) ______________________

(Date) **Issue Date** (Building Official Signature) **Building Official**

The issuance or granting of a permit shall not be construed to be a permit for, or an approval of, any violation of any of the provisions of the city building code or of any other ordinance of the jurisdiction.

A copy of this building permit shall be kept on the site of the work until the completion of the project.

The permit holder or authorized agent is responsible for calling for all required inspections before work is covered and for providing access to the work. Do not occupy this building, or portion of building as described, until final inspection, approval and issuance of the certificate of occupancy.

This permit expires ___/___/___

City Building Department
(000) 000-0000

FIGURE 2-8 Sample of a building permit card

Department of Building Safety
Phone (###) 555-4567

INSPECTION APPROVED

☐ Building   ☐ Plumbing
☐ Electrical ☐ Mechanical

Description: _____

Comments _____

_____

_____

Date: _____

Inspector: _____

FIGURE 2-9 Inspection approval tag

INSPECTIONS

Inspection of the work by qualified personnel at various stages throughout the construction process is essential to verify compliance with the code and the approved plans. It is the responsibility of the permit holder or agent to call for the required inspections before work is concealed and to provide access to such work. If inspection reveals work that does not comply with the code, the inspector notifies the permit holder or agent of the deficiencies requiring correction. The inspector must approve the corrected portions of the work before they are concealed. When work is satisfactory, the inspector typically indicates approval with an inspection sticker or tag (Figure 2-9) or by signing a record-of-inspections card authorizing work to proceed (Figure 2-10). **[Ref. R109]**

Inspection Record

Building Department
City Of

Permit No. _____

Name _____

Address _____

INSPECTOR SHALL SIGN ALL SPACES WHICH APPLY TO THIS JOB

Inspection Category	Date	Comment	Inspector
Foundation Inspection:			
Setbacks, Footings			
Under Slab Inspections:			
Plumbing			
Mechanical			
Electrical			
Utility Inspections:			
Electrical Service			
Gas Piping/Air Test			
Rough-In Inspections:			
Plumbing			
Mechanical (HVAC)			
Electrical			
Framing			

DO NOT COVER WORK
UNTIL IT IS INSPECTED, APPROVED AND ABOVE SPACES ARE SIGNED

Inspection Category	Date	Comment	Inspector
Final Inspection			

DO NOT OCCUPY
Final Inspection Approval and Certificate of Occupancy
issued by the building department are required before occupying this building

FIGURE 2-10 Jobsite inspection record card

Required inspections

The IRC specifically requires certain inspections during the course of construction when applicable, as set forth in Table 2-1. **[Ref. R109]**

TABLE 2-1 Required inspections

Inspection	Conducted when	Inspect
Foundation	• Excavation complete • Forms set • Reinforcing in place • Before any concrete placed	• Setback from lot lines • Dimensions • Reinforcing • Suitable soil or base materials • Vegetation, loose soil and debris removed • Sufficient depth for frost protection • Concrete mix specifications
Floodplain	• The lowest floor is in place (applies to only designated flood hazard areas)	• Elevation certificate prepared and sealed by a registered design professional before work proceeds above
Plumbing, mechanical, gas and electrical rough-in inspections	• Underground work complete • Rough-in stage complete prior to rough frame inspection • Utility connections	• Materials, fittings and methods • Work properly supported and protected • Pressure testing of piping systems
Frame and masonry	• Plumbing, mechanical and electrical rough inspections approved • Framing and masonry complete	• Size, spacing, connection and continuity of all structural members • Load path from roof to foundation • Engineered components • Draftstopping and fireblocking • Stair rise and run • Locations and dimensions of emergency-escape openings

Fire-resistance-rated construction	• Gypsum board fire resistance rated assemblies
	   • Between dwelling units
	   • Exterior walls due to fire separation distance to lot lines
		• Gypsum materials
		• Fastener type, size and spacing conform to the approved fire resistance rated assembly details

Final	• Work under permit complete and ready for occupancy	• Comfort heating, cooling and ventilation equipment and systems • Plumbing fixtures and systems • Electrical circuits, devices and fixtures • Stairs, railings, landings and other means of egress components • Smoke alarms • Emergency-escape and -rescue openings • Other items regulated by the code

FIGURE 2-11 Precast concrete is outside the prescriptive provisions of the IRC but is approved as an engineered system and inspected for conformance to the construction documents.

Other inspections

The building official is authorized to require additional inspections as may be necessary to ensure compliance with the code. Many jurisdictions perform insulation and energy inspections to verify compliance with energy conservation provisions. Inspections may also be desirable to verify the installation of unusual, special, or engineered components or systems. For example, a cast-in-place or precast structural concrete floor or deck (Figure 2-11) is an engineered structure, the design of which is outside the prescriptive provisions of the IRC. The inspector must verify dimensions, materials, anchorage, reinforcing and other details to ensure conformance to the engineered drawings.

Certificate of occupancy

The IRC requires issuance of a certificate of occupancy (Figure 2-12), indicating that the work has passed final inspection before a dwelling unit, or the portion of the dwelling unit covered by the permit can be occupied. **[Ref. R110]**

Certificate of Occupancy

Jurisdiction of _____

This structure or portion of structure as described below has been inspected for compliance with the International Residential Code (IRC) and is hereby issued a certificate of occupancy.

Description of applicable portion of structure _____	Building permit number _____
Address of the structure _____	Code edition **2018 IRC**
Name of owner or owner's authorized agent _____	Sprinkler system required ☐ yes ☐ no
Address of owner or owner's authorized agent _____	Sprinkler system installed ☐ yes ☐ no
Special conditions	_____
Building Official _____	

FIGURE 2-12 Required information for certificate of occupancy

BOARD OF APPEALS

The administration chapter of the IRC creates authority and duties for the building official but intends that actions in enforcing the code be reasonable, and also clearly expresses rights of due process for the public. Such is the case in the right to appeal an order, decision or determination of the building official. Any aggrieved party with a material interest in the decision of the building official may apply for redress to the board of appeals (Figure 2-13). The governing body, such as the city council, appoints the board members, who are qualified by experience and training to hear and rule on matters related to building construction. The intent is to put in place an objective and knowledgeable group of citizens to review the decisions of the building official and consider the merits of any appeal.

Code Essentials

Basis for appeal

- The code has been interpreted incorrectly.
- The code does not apply.
- An equal alternative is proposed.

Building Department
City Of

Department of Building Safety

Application for Appeal

Board of Appeals

Project address ______________________

Use of structure ______________________

Description of work ______________________

Owner's name ______________________ Phone ______________________

Owner's address ______________________

In accordance with the provisions of Section R112 of the *International Residential Code for One- and Two-Family Dwellings* (IRC), I hereby appeal to the Board of Appeals the determination made by the building official relative to the interpretation of Section __________, in order that I might construct the above structure or portion thereof as proposed and shown on the attachments.

Appellant is advised to submit any documentation in support of the appeal. An application for appeal shall be based on a claim that the true intent of the code has been incorrectly interpreted, the provisions of the code do not fully apply, or an equally good or better form of construction is proposed. The board has no authority to waive requirements of the code. Appellant and any interested party may appear to present reasons for granting an appeal at the time of the scheduled meeting.

____
Signature of owner or appellant Date

Meeting date ______________________

Time ______________________

Location ______________________

FIGURE 2-13 Application for appeal

The IRC limits the basis for appeal to matters pertaining to the code requirements. The appellant must claim that the building official has erred in interpreting the code or has wrongly applied a code section. The other basis for appeal is that the appellant considers a proposed alternative to be equal to the code requirements. The IRC does not permit the filing of an appeal seeking a variance or a waiver, and the board has no authority to waive code requirements. **[Ref. R112]**

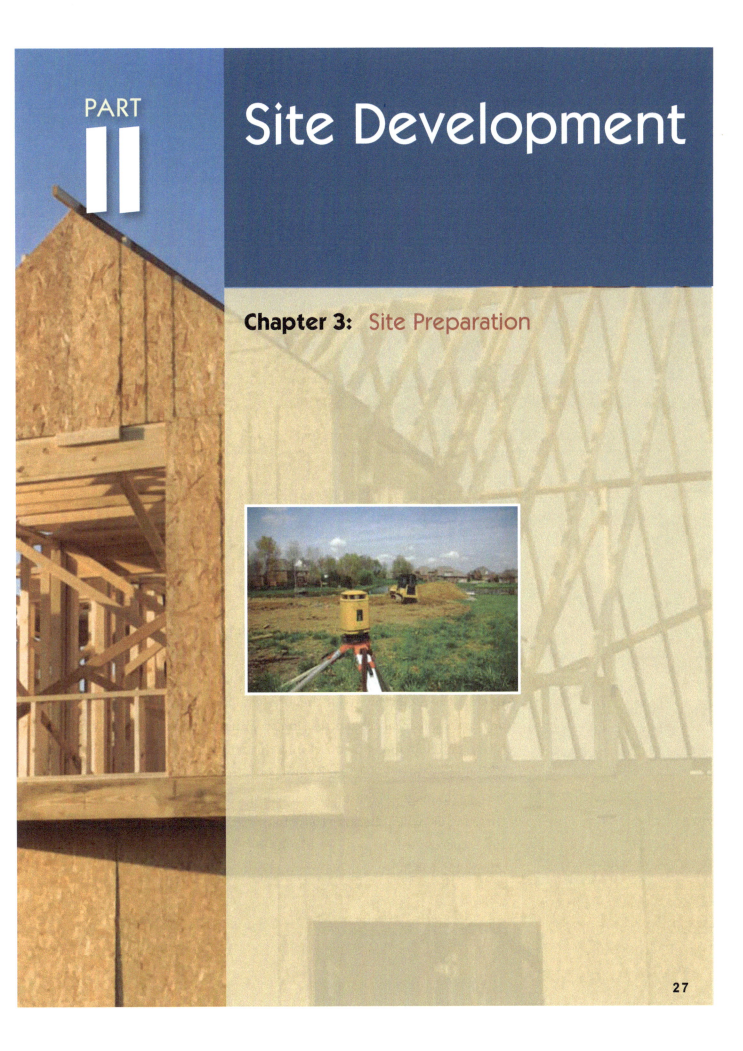

PART

II

Site Development

Chapter 3: Site Preparation

27

CHAPTER 3

Site Preparation

In preparation for constructing buildings on a property, the builder must consider a number of factors related to code requirements. The buildings must be located according to the approved site plan to meet the requirements of the *International Residential Code* (IRC) and any applicable local ordinances. The soil must be suitable for the support of the building and is factored into the design of the foundations. And the building must be elevated sufficiently and the site graded to provide surface drainage away from the building. The plans examiner considers these factors when checking the construction drawings and site plan, but the inspector will be responsible for verifying the requirements at the jobsite (Figure 3-1).

FIGURE 3-1 Sitework

LOCATION ON PROPERTY

The IRC regulates a building's location on the property primarily to guard against the spread of fire. The code is concerned with not only protecting the new building on the property being developed, but preventing the spread of fire to buildings on the adjacent property. Structural considerations also play a part in locating buildings on a lot. The code regulates distances between the structure and adjacent steep slopes to protect the integrity of the foundation. Local zoning or other ordinances may be more restrictive in regulating the location, height and area of buildings on properties.

Fire separation distance

By definition, fire separation distance (FSD) is measured from the face of the building to the lot line, centerline of a street or alley, or to an imaginary line between two buildings. However, for all practical purposes, fire separation distance typically will be of concern only when measured to the interior lot line. No separation distance or fire resistance rating is required for opposing walls of detached dwellings and accessory structures on the same lot. Fire separation distance is measured at a right angle to the face of the exterior wall (Figure 3-2). [Ref. R202]

Provisions that regulate the construction of exterior walls in proximity to lot lines have long been recognized as effective in preventing the spread of fire from a building on one property to a building on another property. Protection can be achieved by providing a clear space between the building and lot line or by using fire-resistant-rated construction. The code does not prohibit placing a building with zero clearance to the lot line provided the exterior wall meets the prescribed fire resistance requirements. When the building is set a certain distance away from the lot line, fire resistance is not required. For dwellings and townhouses protected with an automatic fire sprinkler system, this minimum separation distance is 3 feet. For dwellings without sprinkler systems and for detached accessory buildings, the minimum separation between the unrated wall and the lot line is 5 feet. (See Chapter 9 for further discussion of fire separation distance and fire-resistant protection of exterior walls.) [Ref. R302.1]

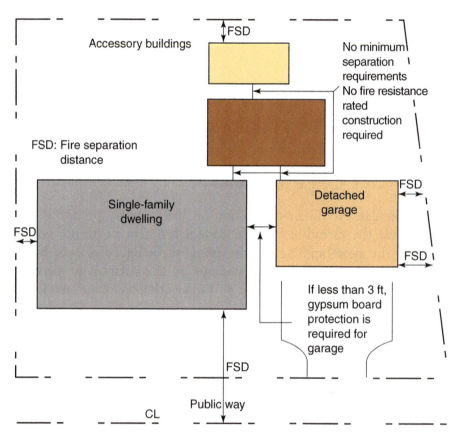

FIGURE 3-2 Measuring fire separation distance

Location of foundations adjacent to slopes

Where slopes are steeper than 33.3 percent (4 inches per foot), foundations must be located a sufficient distance away from the slope to protect the integrity of the structure and provide adequate lateral support to the footing. The clearance distance is based on the height of the slope. For a building located adjacent to the top of the slope (descending), the minimum distance is the height divided by 3, but does not need to exceed 40 feet. For a building located adjacent to the bottom of the slope (ascending), the minimum clearance is the height divided by 2, but does not need to exceed 15 feet. The code gives the building official the authority to approve alternate setbacks with lesser distances to slopes based on a design by a qualified engineer taking all site conditions into consideration (Figure 3-3). [Ref. R403.1.7]

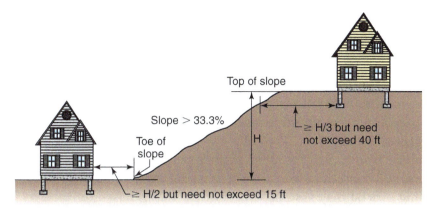

FIGURE 3-3 Foundations adjacent to slopes

SITE PREPARATION

Regulation of site preparation activities related to construction of buildings under the IRC varies based on geographic location and local or site-specific conditions. The code is basically concerned with two things: soil characteristics related to the support and stability of foundations and grading to provide surface drainage away from foundations. Additionally, construction in flood hazard areas must comply with the elevation and design requirements of the IRC or local floodplain regulations. There may also be local or state laws that require grading permits and regulate erosion control, storm water management and soil conservation measures. A number of other factors may affect site preparation and building design, including high water tables and sloped sites.

General site requirements

Preparation of the site for construction includes stripping of vegetation and topsoil, grading to the rough contours if necessary and excavation for basements and foundations. The IRC requires that all exterior footings be placed at least 12 inches below the undisturbed ground level and be protected against frost where applicable. Footings must bear on undisturbed natural soil or compacted engineered fill (covered later in this chapter under "Fill"). The code also prescribes suitable base requirements for basement and garage floors, other slabs on grade and the base for crawl spaces. In all cases, the ground must be stripped of vegetation and organic material. The base for concrete floor slabs within the perimeter walls must be of suitable materials and compacted to prevent settlement. The thickness of compacted fill material below slabs is generally limited to 24 inches for clean sand or gravel and 8 inches for soil unless otherwise approved by the building official (Figure 3-4). [Ref. R403.1, R408.5, R506.2]

FIGURE 3-4 Excavation for foundation of a detached dwelling

Soil properties

The designer or builder must carefully consider soil properties not only for adequate support of the foundation but also for stability to prevent future damage to the structure. Based on experience and known local soil conditions, the building official will often permit design based on a presumptive load-bearing value without soil testing or a geotechnical report. Typically, the presumed load-bearing value will range from 1,500 to 3,000 pounds per square foot (psf) based on local soil conditions and according to the values in Table 3-1. The building official may assume conservative values based on the average or the lowest soil characteristics likely to be encountered on a site. Soil type is verified at the time of footing inspection. If found to

TABLE 3-1 Presumptive load-bearing values and properties of soils

Unified soil classification system symbol	Soil description	Load-bearing pressure (psf)	Drainage	Frost heave potential	Volume change potential expansion
GW	Well-graded gravels, gravel-sand mixtures, little or no fines	3,000	Good	Low	Low
GP	Poorly graded gravels or gravel-sand mixtures, little or no fines	3,000	Good	Low	Low
SW	Well-graded sands, gravelly sands, little or no fines	2,000	Good	Low	Low
SP	Poorly graded sands or gravelly sands, little or no fines	2,000	Good	Low	Low
GM	Silty gravels, gravel-sand-silt mixtures	2,000	Good	Medium	Low
SM	Silty sand, sand-silt mixtures	2,000	Good	Medium	Low
GC	Clayey gravels, gravel-sand-clay mixtures	2,000	Medium	Medium	Low
SC	Clayey sands, sand-clay mixture	2,000	Medium	Medium	Low
ML	Inorganic silts and very fine sands, rock flour, silty or clayey fine sands, or clayey silts with slight plasticity	1,500	Medium	High	Low
CL	Inorganic clays of low to medium plasticity, gravelly clays, sandy clays, silty clays, lean clays	1,500	Medium	Medium	Medium to low
CH	Inorganic clays of high plasticity, fat clays	1,500	Poor	Medium	High
MH	Inorganic silts, micaceous or diatomaceous fine sandy or silty soils, elastic silts	1,500	Poor	High	High

[Ref. Tables R401.4.1 and R405.1]

be of a poorer grade than the presumed value, testing or mitigation is required prior to placing concrete footings. The builder always has the option of providing the results of soil testing in a geotechnical report in order to use a higher load-bearing value than would otherwise be presumed. [Ref. R401.4.1]

Where available data indicates that the soil may not be suitable for the foundation design, the building official is authorized to require a geotechnical evaluation and report prepared by an approved agency or registered design professional. Expansive, compressible or shifting soils have the potential to damage the structure. Highly organic soils (laden with decayed material from plants and animals), such as organic clays, organic silts and peat, are not included in Table 3-1 and are outside the scope of foundation design under the IRC. In addition to organic materials, certain inorganic clays and silts are highly expansive. Such soils expand when wet and contract as they dry, exerting significant pressures against the footing and foundation and thereby causing shifting or differential settlement that could result in structural failure. Expansive soil conditions require an engineered foundation design in accordance with the *International Building Code* (IBC). In some cases it may be possible to remove unsuitable shifting or compressible soils from the building site and replace them with approved fill to stabilize the soil below and around foundations. Under these conditions, the IRC permits a prescriptive foundation design without a full geotechnical evaluation. [Ref. 401.4.2]

Fill

Overexcavation to remove unsuitable soils or the addition of material to raise the elevation of the footings above the level of the natural undisturbed soil requires engineered fill material to support the footings and foundation. A registered design professional is responsible for the design and placement of the fill material in accordance with accepted engineering practice. The engineered fill must be installed and tested in conformance with the design requirements. Fill materials are typically sand, crushed rock, clean gravel or a mix of granular materials. Fill material may contain finer particles that fill voids and help bind the larger elements together. Materials with rounded edges such as river rock or pea gravel are not usually considered suitable for structural fill. The engineer's design specifies the maximum thickness of each layer of fill, called a *lift*, prior to mechanical compaction. A technician tests the compacted fill to verify that it meets the minimum compaction and design specifications. Builders should also exercise care during the backfill of foundations with suitable fill materials to provide adequate drainage and to prevent damage to the foundation. [Ref. R106.1, R401.2]

STORM DRAINAGE

The IRC prescribes methods to direct surface water away from the foundation to an approved location. Water held against the foundation leads to wet or damp basements or crawl spaces and over time can cause damage to construction materials both inside and outside the structure. Mold thrives in such moist environments, contributing to an unhealthy living environment. In addition, water saturation of the soils adjacent to foundations increases the lateral pressure against the structure. Proper design of surface drainage also prevents nuisance ponding on the lot and possible flooding of structures during periods of heavy rain.

The IRC lends some discretion to the building official in determining alternate methods for adequate drainage. Department policy for verifying proper surface drainage on properties will likely vary depending on geographic location, permeability of soils and local history of damage and nuisances created by inadequate drainage. The building official is authorized to require submittal documentation sufficient to demonstrate compliance with the code. If deemed necessary, this may entail a detailed drainage plan with existing and proposed topographic contours, elevations, points of discharge and any containment features. The building official may require that a registered design professional prepare such drainage plans. In many cases, a drainage plan is already established as part of the master plan for the entire housing development and additional plans are not necessary. Other jurisdictions may require only some indication of the direction of drainage flow on the required site plan or may verify drainage on site visually without measurement at the time of inspection (Figure 3-5).

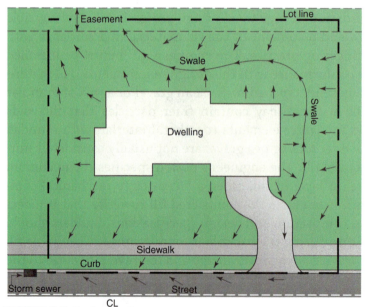

FIGURE 3-5 Drainage plan

The IRC is most concerned with drainage in the immediate vicinity of the structure. The surface of the final grade is required to fall a minimum of 6 inches within the first 10 feet away from the foundation (Figure 3-6). Depending on local site conditions, it is not always possible to achieve that much fall and the code permits alternative designs to drain the water away from the foundation. In this case, the surface water may be directed to swales or drains to ensure adequate drainage away from the structure. Impervious surfaces within 10 feet of the foundation, such as concrete driveways, sidewalks and patios, must be sloped not less than 2 percent away from the structure (Figure 3-7). [Ref. R401.3, R404.1.6]

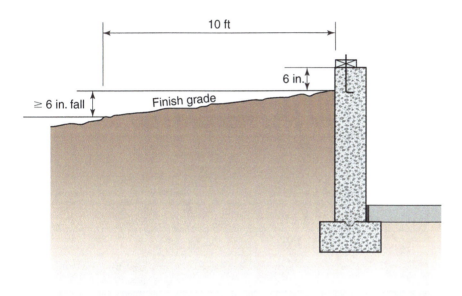

FIGURE 3-6 Grade sloped 6 inches in 10 feet to provide surface drainage away from foundation

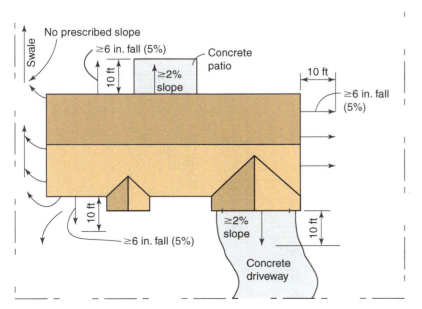

FIGURE 3-7 Grade to ensure surface drainage away from structure

Though the prescribed slopes as previously discussed are concerned with the first 10 feet away from the structure, the IRC also has requirements for drainage to an approved location such as a storm drain, storm sewer inlet or the street gutter that leads to a storm drain. The drainage design must consider the entire lot for any impediments to drainage during heavy rains. **[Ref. R401.3]**

FLOOD HAZARD AREAS

Site development in designated flood hazard areas must meet special requirements for flood resistant construction to minimize damage during a flood event. State or local floodplain management ordinances often supersede the IRC provisions for flood resistant construction. Requirements are similar, however and are designed to satisfy the minimum standards set forth in the National Flood Insurance Program (NFIP), managed by the Federal Emergency Management Agency (FEMA) (Figure 3-8). **[Ref. R322]**

FIGURE 3-8 Flooded dwellings

PART III

Structural

Structural Design Criteria

The *International Residential Code* (IRC) establishes minimum structural design criteria necessary to accommodate normal loads placed on a building and to resist the forces of natural hazards such as wind, snow, earthquake and flood. In most cases, the prescriptive provisions of the code incorporate these criteria and offer a means of conventional construction without the need for an engineered design or complex calculations (Figure 4-1). To correctly apply the values of the tables and the prescriptive methods of construction, it is necessary to understand the structural design criteria based on geographic location and climate. As part of the model code adoption process, the IRC directs the jurisdiction to establish the applicable design criteria and place the corresponding values in Table R301.2(1) (Table 4-1). In addition to structural design considerations, the jurisdiction establishes criteria for environmental hazards such as roof ice dams and termites, as well as local design values for the sizing of heating and cooling systems. These topics are discussed in later chapters. **[Ref. R301.1, R301.2]**

TABLE 4-1 Climatic and geographic design criteria

Ground snow load	Wind design				Seismic design category
	Speed (mph)	Topographic effects	Special wind region	Wind-borne debris zone	

Subject to damage from			Winter design temp	Ice barrier underlayment required	Flood hazard	Air freezing index	Mean annual temp
Weathering	Frost line depth	Termite					

Manual J design criteria						
Elevation	Latitude	Winter heating	Summer cooling	Altitude correction factor	Indoor design temperature	Design temperature cooling

Heating temperature difference	Cooling temperature difference	Wind velocity heating	Wind velocity cooling	Coincident wet bulb	Daily range	Winter humidity	Summer humidity

[Ref. IRC Table R301.2(1)]

PRESCRIPTIVE AND PERFORMANCE

The intent of the code is to provide comprehensive but easy to use minimum standards for the conventional construction of residential buildings and at the same time provide the greatest design flexibility in recognizing other methods and materials of construction. With this in mind, the IRC contains both prescriptive and performance requirements. *Prescriptive* means a set of rules the builder may follow to ensure that the building complies with the code. *Performance* means an expectation that the building system will function in a certain way to meet the minimum requirements of the code. In terms of the structural requirements, performance is typically achieved through engineering.

FIGURE 4-1 Concrete, dimension lumber, engineered wood and steel structural elements of a single-family dwelling

When using the conventional construction provisions, an engineered design is necessary for only those structural elements that exceed the limits of or are otherwise not included in the prescriptive provisions of the code. For example, the sizing of wide flange steel beams commonly used in dwelling construction is outside the scope of the IRC conventional framing systems and must be designed in accordance with accepted engineering practice. This does not prevent the designer and builder from using the prescriptive methods for the rest of the building. In other words, the IRC permits partial or complete engineering of the structure and offers the prescriptive methods as an option, but they are not mandatory. The code imposes seismic, wind and snow loading limitations on the use of the prescriptive framing methods, as will be discussed in later sections of this chapter. [Ref. R301.1.2, R301.1.3]

Code Essentials

Prescriptive vs. performance (structural)

Prescriptive provision:
- Spans for wood floor joists shall be in accordance with Tables R502.3.1(1) and R502.3.1(2)

Performance provision:
- Wood floor trusses shall be designed in accordance with approved engineering practice ●

FIGURE 4-2 *Wood Frame Construction Manual*

FIGURE 4-3 Conventional wood frame construction in progress for a single-family dwelling in accordance with the prescriptive requirements of the IRC

As an alternative to the wood framing provisions of the code, the IRC permits construction to comply with the *Wood Frame Construction Manual* (WFCM), published by the American Wood Council (AWC) (Figure 4-2). The WFCM offers both engineered and prescriptive design requirements for one- and two-family dwellings. The prescriptive design requirements are based on engineering analysis. While the limits of the design provisions are generally consistent with those in the IRC, the WFCM contains provisions for construction in regions with wind speeds that exceed the limits of the IRC prescriptive provisions (Figure 4-3). **[Ref. R301.1.1]**

BASIC LOADS (LIVE AND DEAD)

Building construction must safely support all loads, meaning the forces acting on the building. *Gravity loads* refer to the weight of objects bearing down on the structure and include live loads, dead loads and roof loads. Live loads are the variable loads related to the use of the structure, such as people and furniture. Prescriptive design presumes uniform distribution of the live loads expressed in pounds per square foot (psf) based on the use of the space (Table 4-2). Dead loads are permanent in nature and include the weights of all construction materials and fixed equipment incorporated into the building. The prescriptive tables in the IRC include the combined effects of live loads and dead loads. The total roof load is a combination of dead and live loads, except for buildings in regions where the roof snow load exceeds the roof live load. The code assumes a roof live load—the people, materials and

TABLE 4-2 Minimum uniformly distributed live loads

Use	Live load (psf)	Note
Rooms other than sleeping rooms	40	
Sleeping rooms	30	
Balconies and decks	40	
Stairs	40	Concentrated load of 300 lb. / 4 sq. in. of tread
Guards and handrails		Concentrated load of 200 lb. applied in any direction
Guard in-fill components		Horizontally applied load of 50 lb. on an area of 1 sq. ft.
Passenger vehicle garages	50	Elevated garage floors must support a concentrated load of 2,000 lb. / 20 sq. in.
Attics without storage	10	
Attics with limited storage	20	A storage area that is at least 24 in. wide × 42 in. high with an access hatch or pull-down stair
Attics served by a fixed stair	30	
Habitable attics	30	Floor area ≥ 70 sq. ft. meeting ceiling height requirements

[Ref. Table R301.5]

equipment associated with roofing activities—at not greater than 20 psf. The roof framing is required to support the roof live load or the snow load, whichever is greater. [Ref. R301.4, R301.5, R301.6, Table R301.5]

Live loads

Designs for bedroom areas assume a uniform floor live load of not less than 30 psf, and all other living areas of a dwelling require a minimum live load of 40 psf. Such loads are reflected in the prescriptive tables of the code. Elevated garage floors for vehicles must be designed for a minimum uniform live load of 50 psf and be capable of supporting a concentrated load of 2,000 pounds on any 20-square-inch area. Typically requiring an engineered design, this criterion is necessary to accommodate the concentrated load of a vehicle transferred to the relatively small area of the tires in contact with any portion of the floor.

The performance requirements of stairs and railings present some challenges to the builder and inspector in verifying compliance with the code. Although the IRC does specify minimum loading requirements, it does not include prescriptive structural design provisions for these elements, although numerous conventional and traditional methods are often deemed acceptable without requiring engineering or supporting documentation. Guards and handrails must be secured to safely resist the code-prescribed forces that could act against them. Handrails and top rails of guards are required to be constructed so that they are capable of resisting a 200-pound concentrated force from any direction. The infill components of a guard—often spindles, balusters or intermediate rails—must be able to safely resist 50 pounds applied over a 1-square-foot area.

Although engineering methods could be used to design guard systems and mechanical instruments are available to measure the forces applied to railing components, inspectors more often make a subjective determination of their adequacy based on the perceived strength and rigidity of the components when force is applied against them.

The same may be said for stairs. Stairs are required to be designed for a uniform load of 40 psf, and the treads must be able to support a concentrated load of 300 pounds acting on an area of 4 square inches. This is a conservative value to account for the entire weight of a person on the ball of the foot bearing on the tread. In most cases, $1^1/_4$-inch and 2-inch nominal thickness dimension lumber and manufactured composite stair treads are presumed to accommodate this concentrated load without supporting justification. The builder and inspector often judge the adequacy of the stair to resist the required loads by experience and simply walking the stair. Stiffness or bounce of the stair should be comparable to walking across a floor system that is properly constructed to comply with the code.

Roof live load is a variable load that generally occurs during roofing or roof maintenance operations. The basic roof live load is 20 psf but may be reduced based on a steeper roof pitch or averaging the load effects over a larger area. For other than an engineered design, these roof live load reductions are typically not a factor in construction under the IRC because the prescriptive tables assume a roof live load of 20 psf. For areas with a roof snow load greater than 20 psf, the snow load controls the roof design.

Dead loads

Average dead loads are also included in the prescriptive tables for footings, floors, walls, and roofs. For example, spread footing sizes for conventional frame construction assume average weights of construction materials being supported. Therefore, additional calculations are typically not required. The material and component weights in Tables 4-3 and 4-4 may be helpful in sizing a post or pad footing or another structural element not covered in the IRC tables.

TABLE 4-3 Building material weights

Materials	Weight (psf)
Plywood, ¼ in.	0.8
Plywood, ½ in.	1.6
Plywood, ¾ in.	2.4
Brick, 4 in.	35.0
Gypsum board, ½ in.	2.1
Gypsum board, ⅝ in.	2.5
Plaster, 1 in.	8.0
Stucco, ⅞ in.	10.0
Quarry tile, ½ in.	5.8
Hardwood flooring, ²⁵⁄₃₂ in.	4.0
Built-up roofing	6.5
Shingles, asphalt	1.7–2.8
Shingles, wood	2.0–3.0
Common dimension lumber (pcf)	27–29 pcf
Concrete (pcf)	150 pcf

pcf = pounds per cubic foot

TABLE 4-4 Average weights of building components

Description	Weight (psf)
Roof dead load (framing, sheathing, asphalt shingles, insulation, drywall)	10
Exterior wall (2 × 4 framing, sheathing, siding, insulation, drywall)	10
Floor (joist, sheathing, carpeting, drywall)	10
Concrete wall, 8 in. thick	100
10 in. thick	125
12 in. thick	150
Concrete block wall, 8 in. thick	60

Deflection

Allowable deflection in structural framing members such as studs, joists, and beams is a way to ensure adequate stiffness when such members are subjected to bending under code-prescribed loads. For a floor joist, this may be understood as the bounce or give in the floor system as a person walks across a room. A design for less deflection will translate to more stiffness and therefore less bounce in the floor. The IRC sets limits on the maximum allowable deflection depending on the type of member involved. The code permits greater deflection, for example, in ceiling joists and rafters than in floor joists. These are minimum code requirements, and the homebuilder or homeowner may desire more stiffness and less bounce in a floor than the code would otherwise allow. Although deflection limitations are incorporated into the prescriptive tables, it is important to understand deflection in using the appropriate table for sizing a framing member. Allowable deflection is measured by dividing the span (L) of the member by a prescribed factor, such as 360 for floor joists. Thus the allowable deflection of a floor joist is expressed as L/360 (Table 4-5 and Figure 4-4).
[Ref. R301.7, Table R301.7]

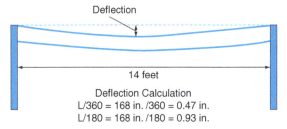

FIGURE 4-4 Deflection limitations for a floor joist and a rafter

TABLE 4-5 Allowable deflection of structural members

Structural member	Allowable deflection
Rafters, slope . ³⁄₁₂, no finished ceiling attached to rafters	L/180
Rafters, slope . ³⁄₁₂, finished ceiling attached to rafters	L/240
Gypsum board ceilings	L/240
Plaster ceilings	L/360
Floors	L/360
All other structural members	L/240
[Ref. Table R301.7]	

Note: Wall deflection and wind load deflections are not shown.

EXAMPLE 4-1

The following example is for a floor joist with a 14-foot span.

$$L = 14 \times 12 \text{ in.} = 168 \text{ in.}$$

Allowable deflection = 168 in./360 = 0.47 in. or approximately ½ inch. Note that a 14-foot span rafter with 4:12 slope and no ceiling attached has an allowable deflection of L/180, which is twice the deflection allowed for floor joists.

WIND, SNOW, SEISMIC AND FLOOD LOADS

In addition to supporting the live and dead loads, the building must safely resist environmental load effects such as wind, snow, earthquake, and flood hazards. These forces may be vertical (up or down) or lateral (sideways) and are also referred to as loads. [Ref. R301.1]

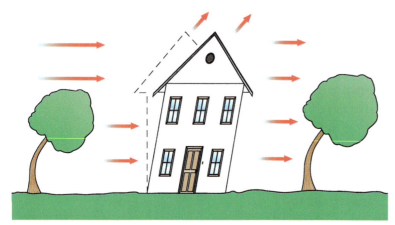

FIGURE 4-5 Wind forces acting on building

Wind

Lateral wind pressure may be positive (pushing against the building on the windward side) or negative (suction forces on the leeward side of a building). Wind pressure can also produce upward suction forces referred to as *uplift*. The building resists wind forces with wall bracing, sheathing, and positive load path connections from the roof down to the foundation (Figure 4-5).

Conventional construction in accordance with the prescriptive provisions of the IRC is generally limited to those geographic regions with wind speeds less than 140 mph as shown in the IRC wind-speed map. Regions where the wind speeds are 140 mph (130 mph for hurricane-prone regions) or greater are identified by the code as requiring wind design, and the prescriptive wind provisions of the code no longer apply (Figure 4-6). In this case, the IRC references several alternatives. Conventional wood frame construction may comply with the

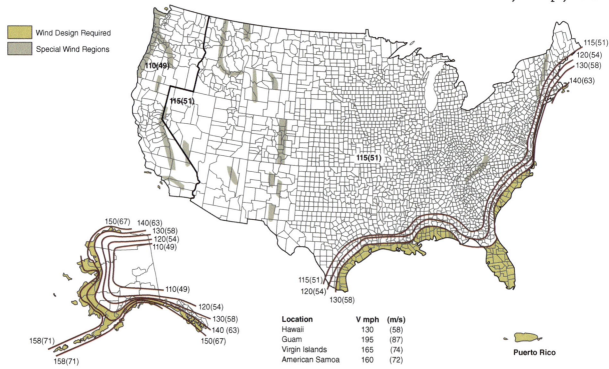

Location	V mph	(m/s)
Hawaii	130	(58)
Guam	195	(87)
Virgin Islands	165	(74)
American Samoa	160	(72)

Notes:
1. Values are ultimate design 3-second gust wind speeds in miles per hour (m/s) at 33 ft (10 m) above ground for Exposure category C.
2. Linear interpolation between contours is permitted.
3. Islands and coastal areas outside the last contour shall use the last wind speed contour of the coastal area.
4. Mountainous terrain, gorges, ocean promontories, and special wind regions shall be examined for unusual wind conditions.
5. Wind speeds correspond to approximately a 7% probability of exceedance in 50 years (Annual Exceedance Probability = 0.00143, MRI = 700 years).

FIGURE 4-6 Regions where wind design is required

WFCM or the ICC Standard for Residential Construction in High-Wind Regions (ICC-600) (Figure 4-7). Otherwise, the code requires wind design in accordance with the engineering provisions of the *International Building Code* (IBC) or ASCE Minimum Design Loads for Buildings and Other Structures (ASCE 7). **[Ref. R301.2.1]**

Exposure category

In addition to the wind speed for a geographic area, ground surface irregularities affect the wind forces placed on a building. Forested terrain or groups of buildings in close proximity shield a building from wind. Flat, open terrain, on the other hand, exposes a building to the full effects of wind.

The IRC classifies wind exposure into three categories: B, C and D. Exposure B, the default and most common application, affords some wind protection with trees and buildings characteristic of urban and suburban settings (Figure 4-8). Exposure C is basically open terrain with scattered obstructions (Figure 4-9). Exposure D applies to buildings in flat, unobstructed areas exposed to wind flowing over open water or mud flats for a distance of 5,000 feet or more.

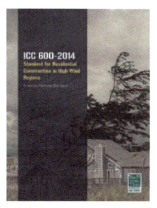

FIGURE 4-7 ICC 600

FIGURE 4-8 Wind exposure B

FIGURE 4-9 Wind exposure C

Exposure categories are important design criteria for engineering of buildings or portions of buildings resisting the effects of wind, and such criteria should appear on engineering submittal documents. Many of the prescriptive methods of wood frame construction in the IRC are generally deemed in compliance without consideration of the wind exposure category. However, wind exposure category is considered when applying the provisions for wall sheathing, wood wall bracing, roof uplift resistance, and exterior wall and roof coverings. Siding, roofing, windows, skylights, exterior doors and overhead doors must be manufactured and installed to resist wind loads based on wind speed, exposure factors and other criteria in accordance with the code. **[Ref. R301.2.1.4]**

You Should Know

Wind design options for wind speed ≥ 140 mph (130 mph hurricane-prone regions)

• WFCM
• ICC-600
• Engineered design (IBC/ASCE 7) ●

Hurricanes

Hurricane-prone regions are the coastal areas of the Atlantic Ocean and Gulf of Mexico where the wind speed is greater than 115 mph. The IRC intends to reasonably limit damage from the destructive forces of hurricanes, even though hurricane winds may significantly exceed the designated wind speeds. The code designates the portions of the hurricane-prone regions that require a design for wind complying with the WFCM, ICC-600, or the IBC. Additional protection is required for exterior glazing in those hurricane areas that are designated by the IRC as windborne debris regions. Typically, glazing protection must meet the referenced testing criteria. However, openings in buildings in regions with a wind speed not greater than 180 mph and a mean roof height of 45 feet or less may be protected with 7/16-inch-thick wood structural panels. The code provides prescriptive methods of attachment for these panels (Figure 4-10). **[Ref. R301.2.1.2]**

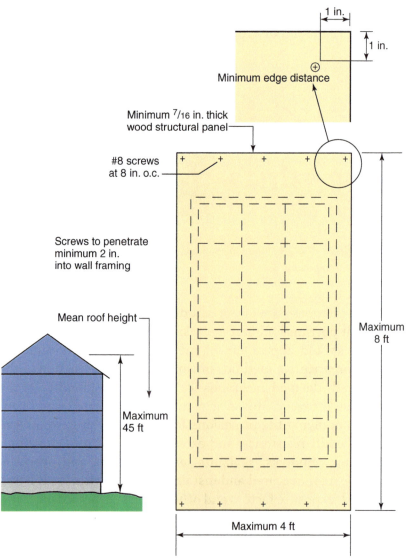

FIGURE 4-10 Prescriptive protection of openings in Wind Zones 1 and 2 of windborne debris regions

Tornadoes

The IRC does not specifically address tornadoes, whose winds may exceed 250 mph, far in excess of design wind speeds. The intent of the code is to provide minimum standards of construction, including those to resist the effects of natural hazards. Buildings constructed in conformance with the code have demonstrated effectiveness in limiting damage due to high winds. The forces and effects of tornadoes are variable and unpredictable. It is not anticipated that a building of wood frame construction in the direct path of a high-intensity tornado will escape without significant damage (Figure 4-11).

FIGURE 4-11 Tornado damage

Storm shelters

Storm shelters, sometimes called safe rooms, though not required by the code, offer added protection from the destructive forces of high winds, hurricanes, and tornadoes. The design and construction of storm shelters, either as detached structures or safe rooms within a dwelling, must conform to the requirements of ICC 500, ICC/NSSA Standard on the Design and Construction of Storm Shelters (Figure 4-12). Primarily designed for life safety considerations, storm shelters protect occupants from serious injury due to high wind and flying debris. The shelters are designed to withstand impact from windborne projectiles, referred to as missiles, such as 2×4s or other construction or natural material debris that are common to high-wind events. The outer shell of above-ground shelters may be concrete, steel, composite or other materials that have been tested to the prescribed large missile tests. [Ref. R323]

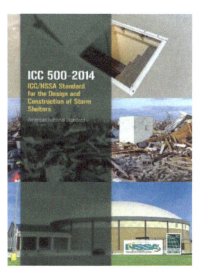

FIGURE 4-12 ICC 500

Snow

The IRC limits the use of the prescriptive provisions to buildings in areas with a ground snow load of 70 psf or less. In areas with higher snow loads, engineering is required for portions or elements supporting a snow load (Figure 4-13). The prescriptive rafter tables reflect ground snow loads of 30, 50 and 70 psf. Engineered components such as roof trusses are designed for the greater of roof live load or roof snow load. Ground snow load is the basis for determining the roof snow load design criteria for engineered components.

FIGURE 4-13 A roof snow load greater than 70 psf is outside the limits of the prescriptive provisions of the IRC.

The IRC snow map specifies the ground snow load for geographic regions. In some regions, indicated by "CS" on the map, and at elevations exceeding the limits shown, the ground snow load must be determined by site-specific case studies. The state or municipal authority may determine the local snow load based upon historical evidence without referencing the map. **[Ref. R301.2.3]**

Earthquake

The design and construction of buildings must resist the load effects caused by ground motion from earthquakes and limit the resulting damage. As indicated in the seismic map in the code, the IRC assigns a seismic design category (SDC) to building sites relative to the anticipated intensity and frequency of earthquakes (Figure 4-14). All

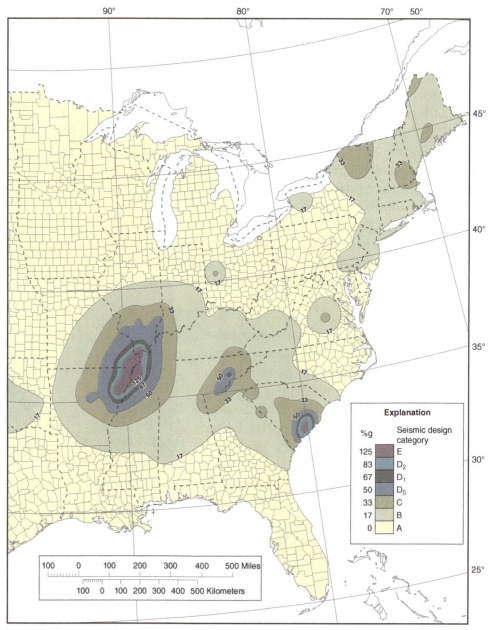

FIGURE 4-14 Seismic design category map

buildings constructed under the prescriptive provisions of the IRC are judged to perform satisfactorily in SDC A and B. Though the conventional methods may be used in SDC A through D_2, specific seismic requirements apply to buildings sited in SDCs C, D_0, D_1 and D_2. However, one- and two-family dwellings are exempt from these additional requirements in SDC C. Buildings in SDC E require an engineered design in accordance with the IBC. **[Ref. R301.2.2]**

Earthquake forces distort the shape of buildings as the foundation moves with the shaking ground motion and the upper portions of the building resist this movement, producing lateral forces that may cause severe damage to the structure (Figure 4-15). This resisting force is referred to as inertia, the tendency of an object at rest to remain at rest. Because inertia forces are directly proportional to the weights of the materials in the upper portions of the building, heavier materials result in higher seismic forces and increased damage. The IRC seismic provisions impose limits on the dead load of the various roof, ceiling, wall and floor elements and may also limit the number of stories. In addition to the lateral, or side-to-side, shaking forces of earthquakes, seismic events may also produce upward forces or change the stiffness characteristics of the soil, resulting in foundation settlement.

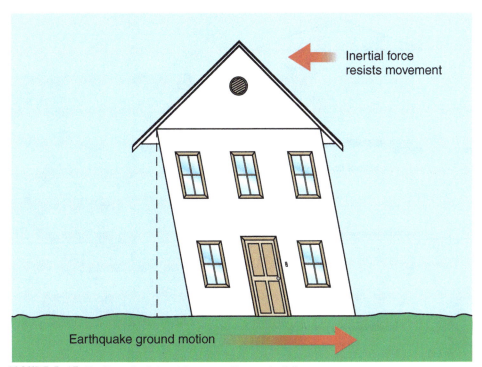

Inertial force resists movement

Earthquake ground motion

FIGURE 4-15 Earthquake lateral forces acting on building

Regular buildings have uniform distribution of forces and more predictable response characteristics when subjected to earthquake ground motion. Irregularly shaped buildings have force concentrations and are generally less effective in resisting earthquake load effects (Figure 4-16). Consequently, except for one- and two-family dwellings in SDC C, the IRC requires engineering for portions of buildings considered irregular in SDCs C, D_0, D_1 and D_2. Among the factors determining irregular buildings are offsets in braced wall lines, arrangement of openings, cantilevers, and the use of dissimilar materials in braced wall lines. **[Ref. R301.2.2.2.5]**

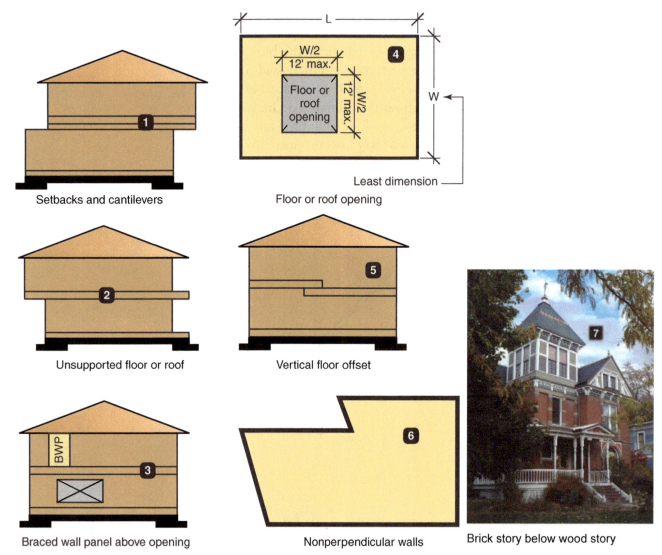

Setbacks and cantilevers

Floor or roof opening

Unsupported floor or roof

Vertical floor offset

Braced wall panel above opening

Nonperpendicular walls

Brick story below wood story

FIGURE 4-16 Seven irregularities requiring engineering in high seismic regions

Floods

The IRC requires flood resistant construction for buildings located in flood hazard areas to minimize damage. Primary protection is achieved through elevation of the lowest floor of the building above the design flood elevation. Basements and spaces used only for storage or vehicle parking are permitted below the design flood elevation. Such enclosed areas require flood openings to allow flood waters to flow through the space and equalize hydrostatic pressure on both sides of the enclosing walls (Figure 4-17). The structural elements of these spaces, as well as building foundations, must effectively resist flood loads in addition to other applicable loads. Flood loads, velocities and hazards associated with floating debris are increased in floodways, the main channel boundaries of a river. Buildings located in a floodway must be designed and constructed in accordance with ASCE 24 to adequately resist these loads. [Ref. R301.2.4, R322]

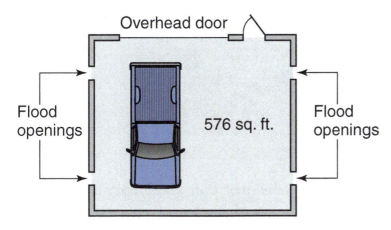

Doors and windows do not satisfy flood opening requirements

Overhead door

Flood openings

576 sq. ft.

Flood openings

Garage below design flood elevation used for parking and storage only

Flood opening requirements

- Two openings required on two sides of the building

- Total 576 sq. in. net area (1 sq. in. per sq. ft. of enclosed area)

- 576/4 = 144 sq. in. net area per opening

- The bottom of each opening no more than 12 in. above grade

FIGURE 4-17 Flood openings

Foundations

The *International Residential Code* (IRC) contains prescriptive foundation designs to safely support building loads and transmit those loads to the soil. This chapter will focus primarily on the provisions for conventional concrete footings and concrete and masonry foundation walls (Figure 5-1). Soil characteristics and bearing capacities affecting foundation design are discussed in Chapter 3.

MATERIALS

The two most common materials for foundation construction are concrete and concrete masonry units (CMUs), the latter often referred to as concrete block or simply masonry. The code does not intend to limit the use of different materials, however. In addition to prescriptive designs for other foundation systems incorporating wood, precast concrete, or insulating concrete forms (ICFs), the IRC permits engineered or alternative designs. [Ref. R402]

Concrete

Concrete continues to gain strength after the initial set through a chemical curing process. The compressive strength of concrete is related to the proportions of portland cement, sand, gravel, and water in the mix and is expressed in pounds per square inch (psi) after 28 days' curing time. The code requires concrete to have a minimum 28-day compressive strength of 2,500 psi for most applications (Table 5-1). Higher-strength concrete, often including entrained air, is specified in geographic areas subject to moderate or severe weathering potential when the concrete is exposed to the weather or is for a garage floor slab (Figure 5-2). [Ref. Table R402.2]

FIGURE 5-1 Concrete foundation forms for a single-family dwelling

TABLE 5-1 Minimum specified compressive strength of concrete

Type or location of concrete construction	Minimum specified compressive strength at 28 Days (psi)		
	Weathering potential		
	Negligible	Moderate	Severe
Basement walls, foundations and other concrete not exposed to the weather	2,500	2,500	2,500[a]
Basement slabs and interior slabs on grade, except garage floor slabs	2,500	2,500	2,500[a]
Basement walls, foundation walls, exterior walls and other vertical concrete work exposed to the weather	2,500	3,000[b]	3,000[b]
Porches, carport slabs and steps exposed to the weather, and garage floor slabs	2,500	3,000[b]	3,500[b]
[Ref. Table R402.2]			

a. Concrete subject to freezing and thawing during construction shall be air entrained.

b. Concrete shall be air entrained.

FIGURE 5-2 Pump to place concrete for the basement of a single-family dwelling

FOOTINGS

In order to properly support the loads of a building, footing design must address not only the size of the footing, but factors such as the condition and characteristics of the soil, footing depth, slope and reinforcing. **[Ref. R403]**

Depth, bearing and slope

For other than engineered soil conditions, footings must bear on undisturbed ground and extend below the frost depth to provide a stable foundation. In addition, exterior footings require excavation to at least 12 inches below the undisturbed soil. Vegetation, wood, debris, loose or frozen soil and any other detrimental materials are removed prior to placing concrete (Figure 5-3). **[Ref. R403.1.4]**

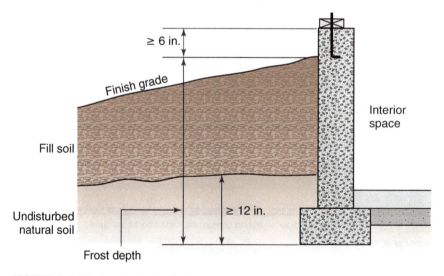

FIGURE 5-3 Depth of exterior footings

Placing footings below the frost depth protects foundations from the expansive effects of freezing and thawing soil. Otherwise, frost heave can exert stresses sufficient to cause significant damage to a foundation. The code offers exceptions to the frost depth requirements where damaging effects of frost heave are negligible. Accessory buildings with an eave height of 10 feet or less and limited in area—600 square feet for light frame construction and 400 square feet for other construction— and decks not attached to a dwelling do not require frost protection. The code also permits frost-protected shallow foundations utilizing rigid polystyrene insulation.

To prevent sliding and to adequately transfer loads to the soil, the code limits the slope of the bottom of footings to a maximum 1 unit vertical in 10 units horizontal (10 percent slope). Transitions that would result in greater slopes must be achieved through stepping of the footings (Figure 5-4). **[Ref. R403.1.5]**

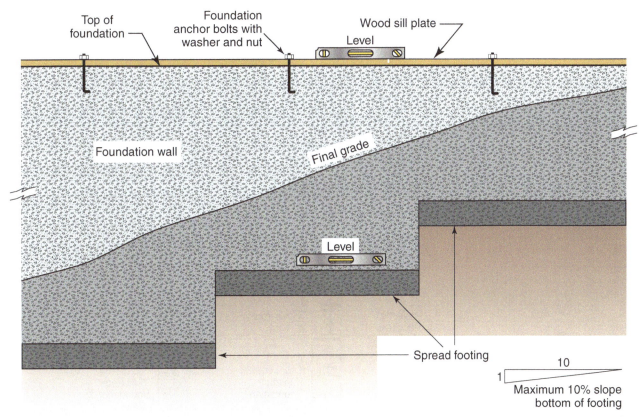

FIGURE 5-4 Stepped footing

Sizing concrete footings

Soil-bearing capacity and the average gravity loads (dead, live, and snow) determine footing size. As the load-bearing capacity of the soil decreases, footing size increases to distribute the load to a greater area. For example, soil with a bearing capacity of 1,500 psf will require a footing with twice the bearing area as soil with a bearing capacity of 3,000 psf when supporting the same total gravity load. Footings must have sufficient bearing area to prevent differential settlement (when the ground settles unevenly), which can cause structural and performance problems. The IRC prescribes the width of continuous footings based on the number of stories supported, the method of construction, the type of foundation (basement, crawl space or slab-on-grade), the snow load and the load-bearing value of the soil (Figure 5-5). The minimum thickness of footings is 6 inches (Table 5-2 and Table 3-2 in Chapter 3). **[Ref. R403.1.1, Tables R403.1(1) through R403.1(3)]**

Width W per Table 5-2
Projection P at least 2 in. and not greater than T
Thickness T not less than 6 in.

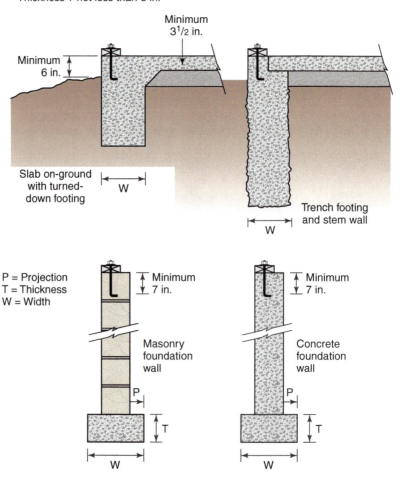

FIGURE 5-5 Types of continuous footings

EXAMPLE 5-1
Size of a continuous footing (width and depth)
Determine minimum width (W), projection (P) and thickness (T) of a continuous spread footing for a two-story dwelling (Figure 5-6) assuming 1,500 psf soil bearing, 30 psf ground snow load and a building width of 32 feet. Use the prescriptive values of Table 5-2. Projection (P) must be at least 2 inches and cannot exceed thickness (T). The minimum footing thickness (T) is 6 inches. The solution is shown in Figure 5-7.

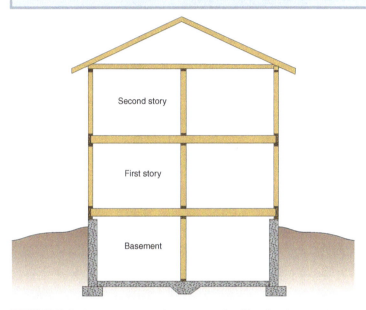

FIGURE 5-6 Cross section of two-story dwelling for determining footing size in Figure 5-7

TABLE 5-2 Minimum width and thickness for concrete footings (inches)

Snow load (psf)	Type of foundation	Load-bearing value of soil (psf)			
30		1,500	2,000	2,500	3,000
Light-frame construction					
1 story	Slab-on-grade	12 x 6	12 x 6	12 x 6	12 x 6
	With crawl space	13 x 6	12 x 6	12 x 6	12 x 6
	Plus basement	19 x 6	14 x 6	12 x 6	12 x 6
2 story	Slab-on-grade	12 x 6	12 x 6	12 x 6	12 x 6
	With crawl space	17 x 6	13 x 6	12 x 6	12 x 6
	Plus basement	23 x 6	17 x 6	14 x 6	12 x 6
Light-frame construction with brick veneer					
1 story	Slab-on-grade	12 x 6	12 x 6	12 x 6	12 x 6
	With crawl space	16 x 6	12 x 6	12 x 6	12 x 6
	Plus basement	22 x 6	16 x 6	13 x 6	12 x 6
2 story	Slab-on-grade	16 x 6	12 x 6	12 x 6	12 x 6
	With crawl space	22 x 6	16 x 6	13 x 6	12 x 6
	Plus basement	27 x 9	21 x 6	16 x 6	14 x 6

[Ref. excerpt from Tables R403.1(1) and R403.1(2)]

Note: Values are based on a 32-foot wide house with load bearing center wall that carries half of the tributary attic and floor framing loads. For every 2 feet of adjustment to the width of the house, add or subtract 2 inches of footing width and 1 inch of footing thickness (but not less than 6 inches thick).

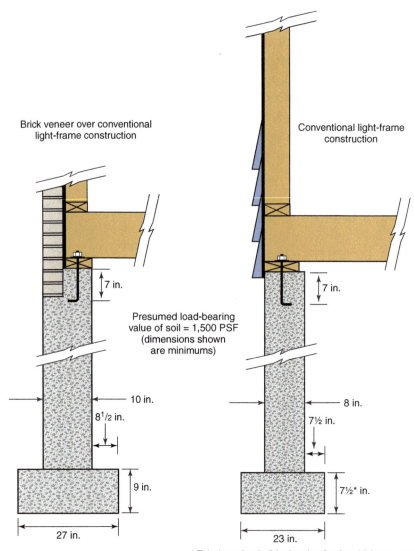

Brick veneer over conventional
light-frame construction

Conventional light-frame
construction

7 in.

7 in.

Presumed load-bearing
value of soil = 1,500 PSF
(dimensions shown
are minimums)

10 in.

8¹/₂ in.

8 in.

7½ in.

9 in.

7½* in.

27 in.

23 in.

* Tabular value is 6 inches but footing thickness cannot be less
than the projection of 7½ inches (23 - 8 = 15 / 2 = 7.5).

FIGURE 5-7 Minimum size of continuous spread footing for a two-story house with
basement based on Table 5-2 and Figure 5-6

EXAMPLE 5-2
Size of isolated footing for a column load
The soil type and the total tributary load being supported determine pier and column footing size. See Figure 5-8 for a simple example of tributary floor load transferred to a column. Assuming a total floor load of 50 psf (40 psf live load plus 10 psf dead load), a tributary area of 120 square feet will result in a column load of 6,000 pounds (120 × 50 = 6,000). A post carrying a 6,000-pound load will require a footing area of 4 square feet (2 feet by 2 feet) for a soil-bearing capacity of 1,500 psf (6,000/1,500 = 4). Minimum footing dimensions are shown in Figure 5-9.

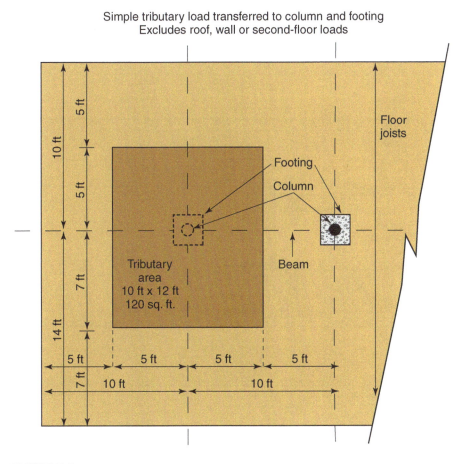

Simple tributary load transferred to column and footing
Excludes roof, wall or second-floor loads

FIGURE 5-8 Tributary load

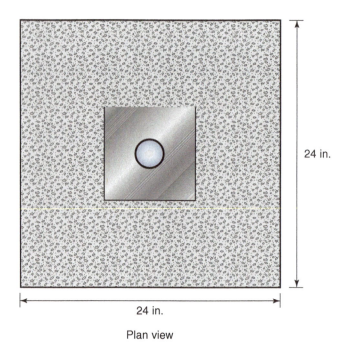

Plan view

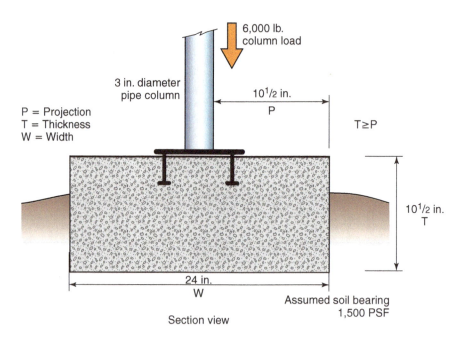

Section view

FIGURE 5-9 Minimum dimensions of an isolated column footing based on tributary load and assumed soil-bearing capacity

Reinforcing for footings

For the most part, the IRC permits footings without reinforcement in seismic design categories (SDCs) A, B and C (Figure 5-10). The code does require reinforcing of footings when the building is located in SDC D_0, D_1 or D_2. See Figures 5-11 and 5-12 for the reinforcing requirements in these higher seismic regions. In addition, some alternate methods for narrow wall bracing require reinforcing in footings and foundations at the braced wall panel locations regardless of the seismic design category. Further information is provided in the discussion on wall bracing in Chapter 6 of this publication. **[Ref. R403.1.3]**

FIGURE 5-10 Concrete spread footing

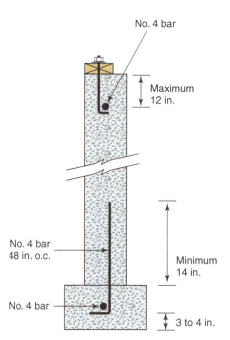

FIGURE 5-11 Concrete footing and stem wall reinforcing in SDCs D_0, D_1 and D_2

Foundation anchorage

Anchorage to the foundation is a critical part of the load path to resist lateral and uplift forces acting on the framing system of the building. The IRC prescribes anchor bolt criteria for connecting the sill plate to the foundation. Other methods, such as foundation straps, may be used if installed according to the manufacturer's instructions and in a way to provide equivalent anchorage. Such alternatives typically require closer spacing than for embedded anchor bolts (Figures 5-13 through 5-15). **[Ref. R403.1.6]**

To resist the increased forces of earthquake ground motion, additional anchorage requirements apply to buildings located in SDCs D_0, D_1 and D_2 and to townhouses in SDC C. In this case, anchor bolts in braced wall lines (described in Chapter 6) require 3-inch by 3-inch plate washers approximately ¼ inch thick to increase the bearing area of the washer against the sill plate or bottom plate so the anchor bolts do not tear out in an earthquake (Figure 5-16). Bolt spacing is also reduced to 4 feet for anchorage of three-story buildings. One- and two-family dwellings are exempt from specific seismic design requirements in SDC C (Table 5-3). **[Ref. R403.1.6.1, R602.11.1]**

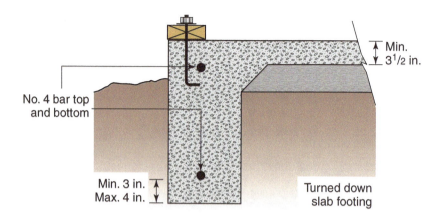

No. 4 bar top and bottom

Min. 3½ in.

Min. 3 in.
Max. 4 in.

Turned down slab footing

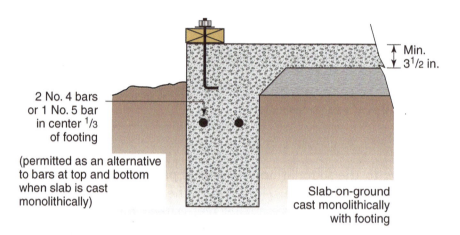

2 No. 4 bars or 1 No. 5 bar in center ⅓ of footing

(permitted as an alternative to bars at top and bottom when slab is cast monolithically)

Min. 3½ in.

Slab-on-ground cast monolithically with footing

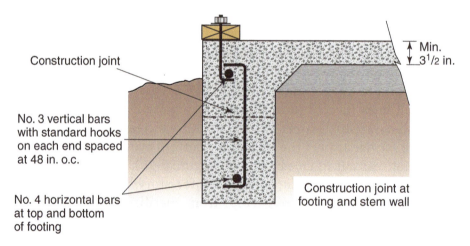

Construction joint

No. 3 vertical bars with standard hooks on each end spaced at 48 in. o.c.

No. 4 horizontal bars at top and bottom of footing

Min. 3½ in.

Construction joint at footing and stem wall

FIGURE 5-12 Reinforcing for turned down slab footings in SDCs D_0, D_1 and D_2

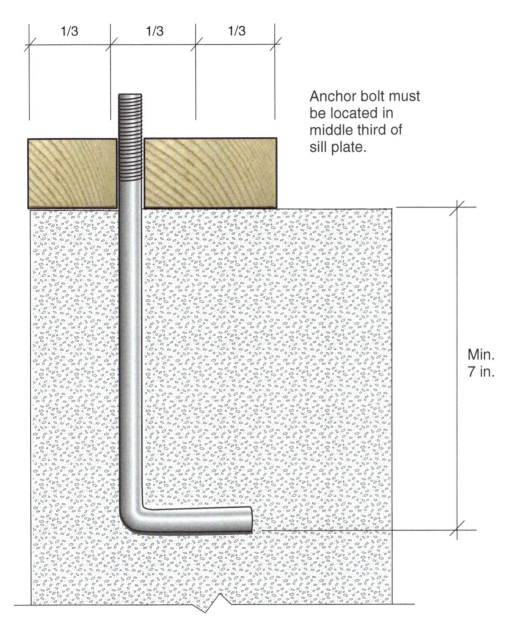

1/3   1/3   1/3

Anchor bolt must
be located in
middle third of
sill plate.

Min.
7 in.

FIGURE 5-13 Anchor bolt placement in center third of sill plate

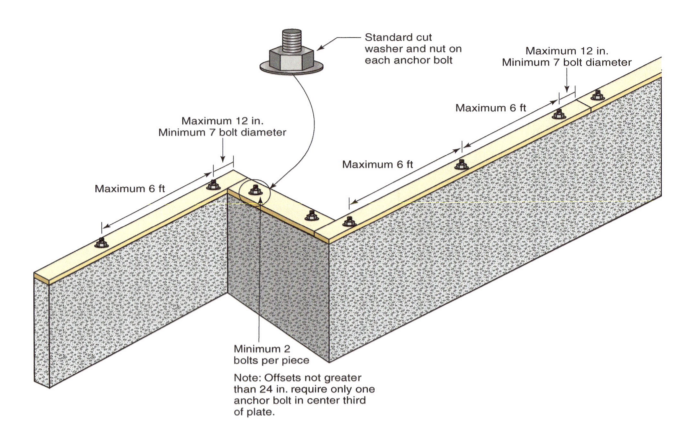

Standard cut
washer and nut on
each anchor bolt

Maximum 12 in.
Minimum 7 bolt diameter

Maximum 6 ft

Maximum 12 in.
Minimum 7 bolt diameter

Maximum 6 ft

Maximum 6 ft

Minimum 2
bolts per piece

Note: Offsets not greater
than 24 in. require only one
anchor bolt in center third
of plate.

FIGURE 5-14 Wood sill plate anchorage to foundation for all buildings in SDCs A and B, and dwellings in SDC C

FIGURE 5-15 Anchor bolts for all buildings in SDCs A and B, and
dwellings in SDC C

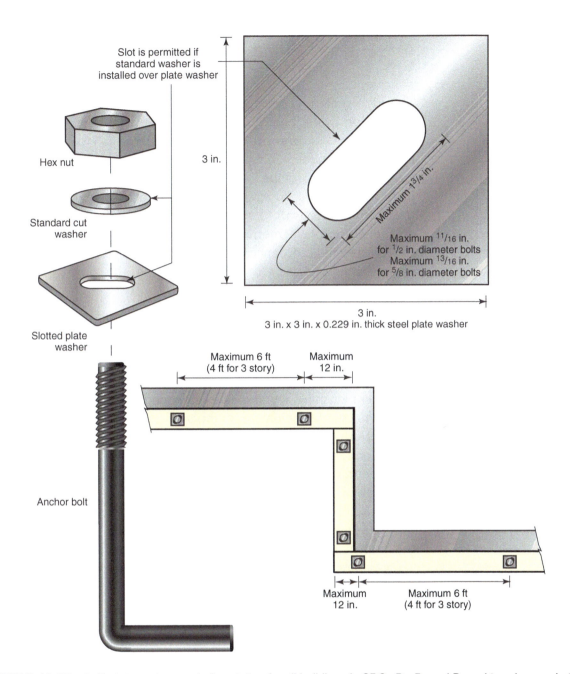

FIGURE 5-16 Wood sill plate anchorage to foundation for all buildings in SDCs D_0, D_1 and D_2 and townhouses in SDC C

TABLE 5-3 Foundation anchor bolts (minimum ½-inch diameter and 7-inch embedment)

Seismic Design Category	One- and two-family dwellings			Seismic Design Category	Townhouses		
	Stories	Maximum spacing	Washer		Stories	Maximum spacing	Washer
A and B				**A and B**	1, 2, or 3	6 ft	Standard cut
C	1, 2, or 3	6 ft	Standard cut	**C**	1 or 2	6 ft	3 in. × 3 in. × 0.229 in. min.
					3	4 ft	
D_0, D_1 and D_2	1 or 2	6 ft	3 in. × 3 in. × 0.229 in. min.	**D_0, D_1 and D_2**	1 or 2	6 ft	
	3	4 ft			3	4 ft	

Note: Standard cut washers are permitted in wall lines without braced wall panels.

MASONRY AND CONCRETE FOUNDATION WALLS

The IRC includes prescriptive methods for the construction of foundation walls that are supported at their top and bottom. Top support to prevent lateral movement is typically achieved through adequate connections to the floor system. Bottom lateral support is provided by the footing in the ground or by the placement of the concrete basement slab floor directly against the foundation wall. Foundation walls that exceed the limits of the prescriptive methods in the code must be designed and constructed in accordance with the referenced standards or accepted engineering practices. [Ref. R404.1]

Wall height and thickness

Unlike footings, where gravity loads are the primary consideration, foundation walls must be constructed to resist lateral loads, particularly from soil pressure. Therefore, the soil type, height of backfill and height of the foundation determine the wall thickness and reinforcement of masonry and concrete foundation walls without consideration of the height or number of stories of the dwelling. Soil descriptions and types are given in Table 3-1 in Chapter 3.

When required, the location, size and spacing of vertical reinforcing depend on the minimum yield strength (grade) of the steel, the thickness of the wall, and other variables. As wall thickness increases, the amount of vertical reinforcing required decreases. Said another way, to reduce the thickness of a foundation wall, say from 12 inches to 8 inches, larger-diameter reinforcing bars or closer spacing of the vertical bars, or both, may be necessary. For concrete basement walls constructed under the prescriptive provisions, 2 or 3 rows of horizontal reinforcing are required based on the unsupported height of the wall [Tables R404.1.2(1) through R404.1.2(9)].

You Should Know

Alternatives for construction of concrete foundations

- IRC
- PCA 100
- ACI 332
- ACI 318
- Engineered ●

EXAMPLE 5-3

Determine the minimum thickness and reinforcing requirements for the concrete basement foundation walls in Figures 5-17 and 5-18 based on Tables 5-4 and 5-5.

TABLE 5-4 Minimum horizontal reinforcement for concrete basement walls

Minimum unsupported height of basement wall (feet)	Location of horizontal reinforcement
≤ 8	One No. 4 bar within 12 inches of the top of the wall story and one No. 4 bar near mid-height of the wall story
> 8	One No. 4 bar within 12 inches of the top of the wall story and one No. 4 bar near one-third points in the wall story

[Ref. Table R404.1.2(1)]

TABLE 5-5 Minimum vertical reinforcement for concrete basement walls

Maximum wall height (feet)	Maximum unbalanced backfill height (feet)	Minimum vertical reinforcement—Bar size and spacing (inches)											
		Soil classes and lateral soil load (psf per foot of depth)											
		GW, GP, SW and SP				GM, GC, SM, SM-SC and ML				SC, ML-CL and inorganic CL			
		30				45				60			
		Minimum nominal wall thickness (inches)											
		6	8	10	12	6	8	10	12	6	8	10	12
9	6	4 @ 34	NR	NR	NR	6 @ 48	NR	NR	NR	6 @ 36	6 @ 39	NR	NR
	7	5 @ 36	NR	NR	NR	6 @ 34	5 @ 37	NR	NR	6 @ 33	6 @ 38	5 @ 37	NR
	8	6 @ 38	5 @ 41	NR	NR	6 @ 33	6 @ 38	5 @ 37	NR	6 @ 24	6 @ 29	6 @ 39	4 @ 48
	9	6 @ 34	6 @ 46	NR	NR	6 @ 26	6 @ 30	6 @ 41	NR	6 @ 19	6 @ 23	6 @ 30	6 @ 39

[Ref. excerpt of Table R404.1.2(8)]

NR = Not required

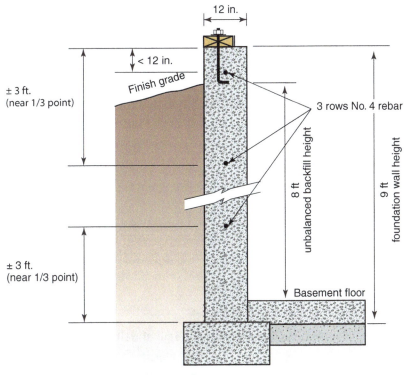

FIGURE 5-17 Wall thickness and horizontal reinforcing requirements for concrete basement wall with no vertical reinforcing for all buildings in SDC A and B, and dwellings in SDC C, based on Tables 5-4 and 5-5

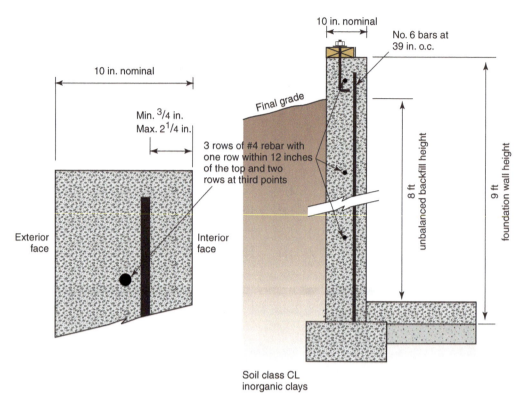

FIGURE 5-18 Wall thickness and reinforcing requirements for flat concrete basement wall with vertical reinforcing based on Tables 5-4 and 5-5

Two examples for determining the thickness of a masonry foundation wall are given in Table 5-6 and Figure 5-19, and Table 5-7 and Figure 5-20. **[Ref. R404.1.2, Tables R404.1.1(1) through R404.1.1(4)]**

EXAMPLE 5-4

Determine the minimum thickness of the masonry basement foundation wall without reinforcing in Figure 5-19 based on Table 5-6.

TABLE 5-6 Plain masonry foundation walls

Plain masonry walls				
		Plain masonry minimum nominal wall thickness (inches)		
		Soil classes		
Maximum wall height (feet)	**Maximum unbalanced backfill height (feet)**	**GW, GP, SW and SP soils**	**GM, GC, SM, SM-SC and ML soils**	**SC, ML-CL and inorganic CL soils**
8	6	10	12	12 fully grouted or solid masonry
	7	12	12 fully grouted or solid masonry	Reinforcing or design required
	8	10 fully grouted masonry	12 fully grouted masonry	Reinforcing or design required

[Ref. excerpt of Table R404.1.1(1)]

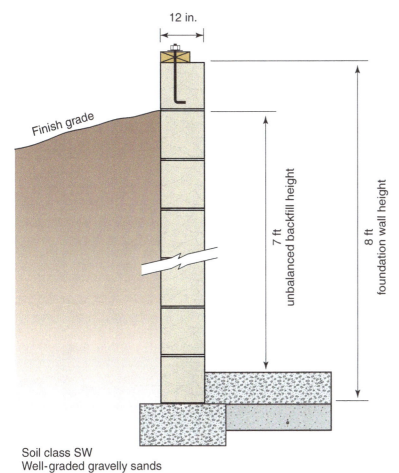

Soil class SW
Well-graded gravelly sands

FIGURE 5-19 Wall thickness for plain masonry (no reinforcing) foundation wall for all buildings in SDCs A and B, and dwellings in SDC C based on Table 5-6

EXAMPLE 5-5

Determine the minimum thickness and reinforcing requirements for the masonry basement foundation wall in Figure 5-20 based on Table 5-7.

TABLE 5-7 Masonry foundation walls with reinforcing

Eight-inch masonry foundation walls with reinforcing where d ≥ 5 inches				
		Minimum vertical reinforcement		
		Soil classes and lateral soil load (psf per foot below grade)		
		GW, GP, SW and SP soils	GM, GC, SM, SM-SC and ML soils	SC, ML-CL and inorganic CL soils
Wall height	Unbalanced backfill height	30	45	60
8 feet 8 inches	6 feet	#4 at 48" o.c.	#5 at 48" o.c.	#6 at 48" o.c.
	7 feet	#5 at 48" o.c.	#6 at 48" o.c.	#6 at 40" o.c.
	8 feet 8 inches	#6 at 48" o.c.	#6 at 32" o.c.	#6 at 24" o.c.

[Ref. excerpt of Table R404.1.1(2)]

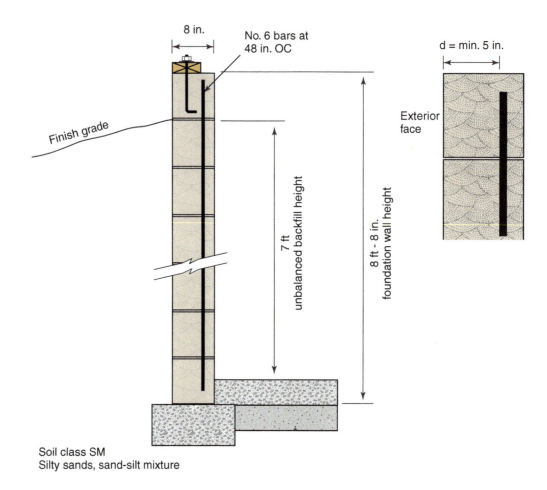

8 in.

No. 6 bars at
48 in. OC

d = min. 5 in.

Finish grade

Exterior
face

7 ft
unbalanced backfill height

8 ft - 8 in.
foundation wall height

Soil class SM
Silty sands, sand-silt mixture

FIGURE 5-20 Wall thickness and reinforcing requirements for masonry foundation wall with reinforcing based on Table 5-7

Seismic requirements

Because masonry and concrete foundation walls without reinforcing are more susceptible to damage from earthquake ground motion than walls with reinforcing, the IRC places additional limitations on unreinforced foundation walls in SDCs D_0, D_1 and D_2. In general, the code limits the wall height to 8 feet, with a maximum unbalanced backfill height of 4 feet and a minimum nominal wall thickness of 8 inches. In addition, where the table allows plain concrete walls, the code requires one vertical No. 4 bar every 48 inches on center. This does not affect the tabular values for concrete and masonry walls with reinforcing. **[Ref. R404.1.4]**

Height above finished grade

The IRC intends to protect the building from moisture intrusion that may damage portions of the structure and cause an unhealthy living environment. As part of these requirements, concrete and masonry foundation walls must extend above the finished grade a minimum of 6 inches. Masonry veneer provides somewhat better protection against moisture intrusion at ground level than do siding and other exterior wall finishes, and the elevation requirement for the top of the foundation wall is reduced accordingly to 4 inches above finished grade (Figure 5-21). **[Ref. R404.1.6]**

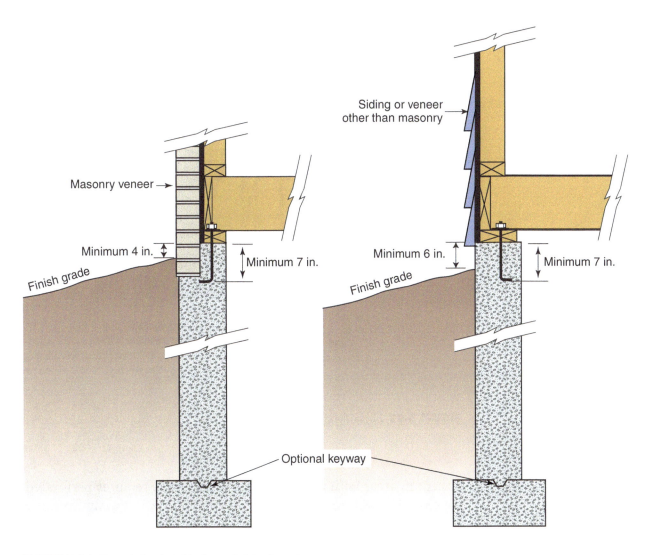

FIGURE 5-21 Foundation height above finished grade

MOISTURE PROTECTION

For foundations that retain earth and enclose spaces located below grade, the IRC requires foundation drainage and dampproofing or waterproofing to prevent moisture from penetrating into such spaces (Figure 5-22).

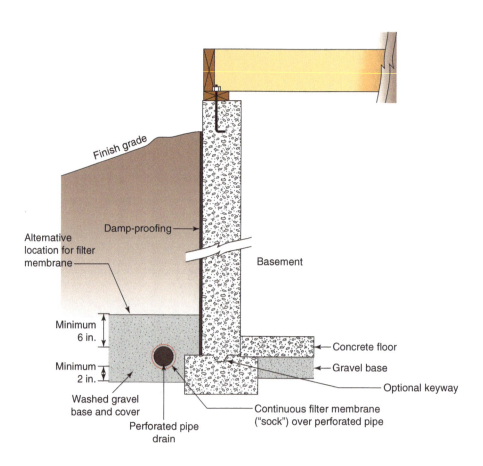

FIGURE 5-22 Foundation drain and dampproofing

Foundation drainage

Except in areas with well-drained soils, the code requires perforated pipe or other approved drains to carry groundwater away from the foundation. Such drains must be installed at or below the level of the basement or crawl space floor using approved methods and the prescribed amounts of washed gravel. Perforated drains require installation of a filter membrane unless the manufacturer recommends otherwise. The drainage system is required to discharge by gravity or mechanical means to an approved location. [Ref. R405]

Dampproofing and waterproofing

Under normal conditions and in combination with foundation drains, dampproofing is deemed adequate to prevent moisture from migrating into the enclosed space. A bituminous-based coating or other approved dampproofing materials are applied to the exterior of the foundation, typically from the top of the footing to the finished grade. Areas with a high water table or other known severe soil-water conditions require waterproofing. Typically consisting of flexible sealants or other impervious material and applied in thicker coatings, waterproofing provides a higher level of protection against moisture under hydrostatic pressure. [Ref. R406]

UNDERFLOOR SPACE

Depending on climatic conditions, significant amounts of condensation can accumulate in enclosed crawl spaces, causing decay and other damage to the structure. The code requires ventilation openings through the foundation or exterior walls in the prescribed size and location to circulate air and dissipate condensation (Figure 5-23).

An alternative method permits a crawl space without foundation openings when equipped with mechanical exhaust ventilation or connection to the conditioned air supply of the dwelling. In this case, the code requires insulation of the exterior walls and a vapor retarder over the ground and sealed to the enclosing foundation wall.

The IRC requires access to the underfloor spaces. Access openings through the floor must be at least 18 inches by 24 inches but may be reduced to not less than 16 inches by 24 inches when access is through a perimeter wall. [Ref. R408]

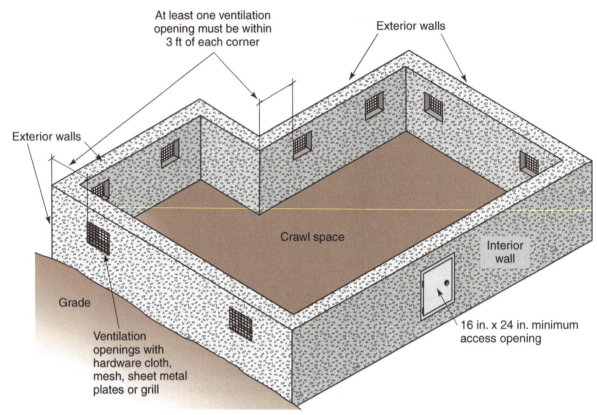

At least one ventilation opening must be within 3 ft of each corner

Exterior walls

Exterior walls

Crawl space

Grade

Interior wall

Ventilation openings with hardware cloth, mesh, sheet metal plates or grill

16 in. x 24 in. minimum access opening

FIGURE 5-23 Crawl space ventilation and access

CHAPTER 6

Framing

The repetitive system of wood or cold-formed steel framing members forming the structural elements of floor, wall and roof construction is referred to as light-frame construction (Figure 6-1). The framing system, with its connections and bracing, must resist the code-prescribed vertical and lateral forces that act on the building. These loads must be adequately transferred through the framing system by a complete load path to the foundation. The *International Residential Code* (IRC) prescribes specific framing requirements that when followed preclude the need for an engineered design. This chapter will focus on the prescriptive provisions for conventional wood light-frame construction.

FIGURE 6-1 Wood framing

GRADE MARKS

Load-bearing dimension lumber for framing members and wood structural panels must be identified by a grade mark (Figure 6-2). Sawn lumber grade marks indicate the wood species, grade, moisture content, grading agency and lumber mill identification. Species and grade determine, in part, the strength and stiffness properties that establish the maximum permissible spans for wood beams, joists and rafters. Wood structural panel grade marks include the maximum span ratings for roof and floor applications to meet minimum performance requirements. **[Ref. R502.1, R503.2, R602.1, R803.1, R803.2]**

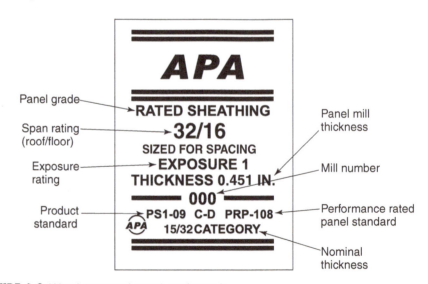

FIGURE 6-2 Wood structural panel grade mark

ENGINEERED WOOD PRODUCTS

Engineered wood products include plate-connected open web trusses, I-joists, glued laminated lumber, laminated veneer lumber and other structural composite lumber. The IRC permits the use of engineered components in otherwise prescriptive conventional framing systems. These engineered components must be designed in accordance with accepted engineering practice and the applicable referenced standards. Installation of engineered wood products must conform to the manufacturer's installation instructions (Figure 6-3).

FIGURE 6-3 Engineered wood products: I-joists and LVL beam

TRUSSES

In addition to structural design criteria, truss design drawings include manufacturing and installation specifications for each truss (Figures 6-4 and 6-5). The IRC requires the manufacturer or contractor to submit the truss design drawings to the building official for review and approval prior to truss installation. Because they contain permanent bracing details, nailing specifications for bracing and multiple-member trusses, and minimum bearing and other important installation information, the truss design drawings must also be delivered to the jobsite with the trusses. The floor or roof sheathing and gypsum board ceiling materials typically provide adequate lateral support and thus satisfy the permanent bracing requirements for the top and bottom chords of the truss. In order to adequately resist the design loads and resist buckling, truss web members in compression often require additional lateral support that is provided by the specified permanent bracing. Alterations to trusses are not allowed without the approval of a registered design professional. [Ref. R502.11, R802.10]

FIGURE 6-4 Open-web floor trusses

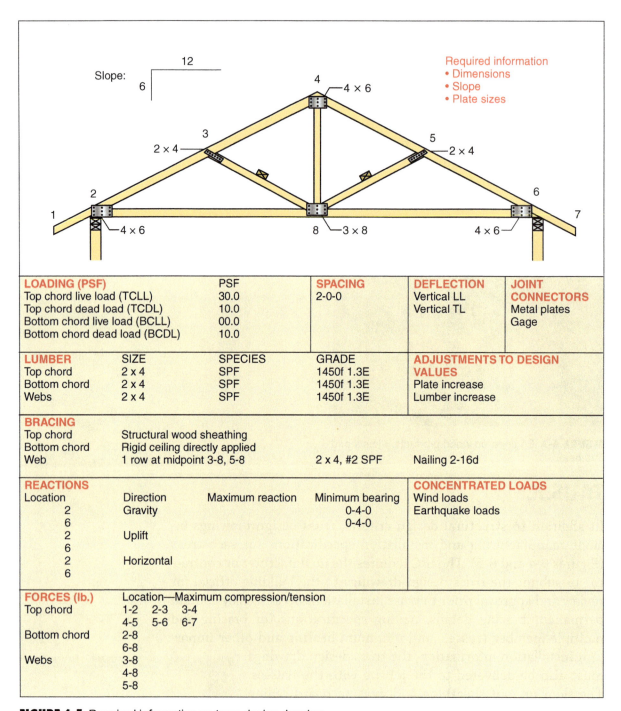

FIGURE 6-5 Required information on truss design drawing

WOOD TREATMENT

Portions of wood construction in locations subject to decay require naturally durable wood or wood treated with preservatives. The heartwoods of decay-resistant redwood, cedar, black locust and black walnut are considered naturally durable. An approved quality mark or label is required on preservative-treated lumber and plywood, indicating that the products meet the standards of the American Wood Protection Association (AWPA) (Figure 6-6). Preservative-treated wood suitable for ground contact is required for structural supports that are in contact with the ground, embedded in concrete in contact with the ground, or embedded in concrete exposed to the weather. Naturally durable wood is not permitted in these ground contact locations. Some chemicals used in the preservative treatment process are corrosive to steel. To resist corrosion and maintain structural load capacity, the code generally requires fasteners and connectors used in preservative and fire-retardant treated wood to be hot-dipped, zinc-coated galvanized steel, stainless steel, silicon bronze, copper or material recommended by the manufacturer (Figures 6-7 and 6-8). [Ref. R317]

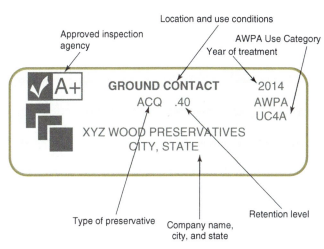

FIGURE 6-6 Example of information on end tag of preservative-treated wood

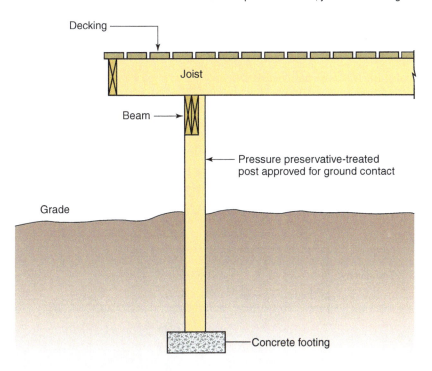

FIGURE 6-7 Protection against decay for wood decks

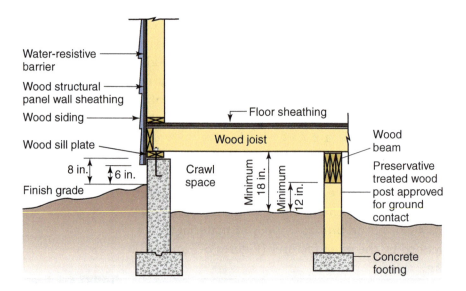

FIGURE 6-8 Clearance above ground for protection against decay

CUTTING, BORING AND NOTCHING

In order to maintain the structural strength and integrity of wood framing, the code limits the amount and location of bored holes and notches in dimension lumber, as shown in Figures 6-9 through 6-12. Fractions and percentages related to the depth of the framing member are converted to inches in Table 6-1. The IRC generally prohibits boring, cutting or notching of trusses and other engineered wood products except as specifically permitted by the manufacturer. Otherwise, a registered design professional must consider any such alterations in the design of the engineered component. **[Ref. R502.8, R602.6, R802.7]**

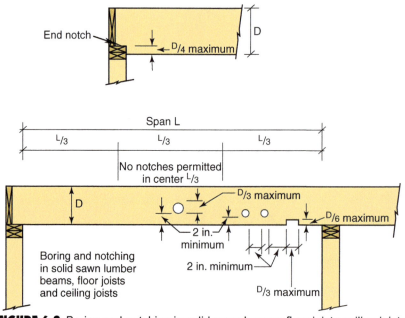

FIGURE 6-9 Boring and notching in solid sawn beams, floor joists, ceiling joists and rafters

TABLE 6-1 Boring and notching limits for wood beams, joists, rafters and studs converted to inches

Sawn lumber beams, floor joists, ceiling joists and rafters		Approximate notch and hole limitations in inches		
Nominal size	Approximate depth 'D'	D/3	D/4	D/6
2 x 4	3 ½ in.	1 $^3/_{16}$	$^7/_8$	½
2 x 6	5 ½ in.	1 $^{13}/_{16}$	1 $^3/_8$	$^{15}/_{16}$
2 x 8	7 ¼ in.	2 $^3/_8$	1 $^{13}/_{16}$	1 $^3/_{16}$
2 x 10	9 ¼ in.	3 $^1/_{16}$	2 $^5/_{16}$	1 ½
2 x 12	11 ¼ in.	3 ¾	2 $^{13}/_{16}$	1 $^7/_8$
Wood studs		**60%**	**40%**	**25%**
2 x 4	3 ½ in.	2 $^1/_8$	1 $^3/_8$	$^7/_8$
2 x 6	5 ½ in.	3 $^5/_{16}$	2 $^3/_{16}$	1 $^3/_8$

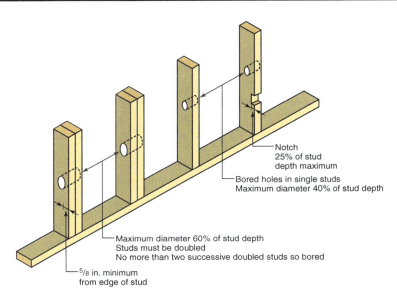

Notch
25% of stud
depth maximum

Bored holes in single studs
Maximum diameter 40% of stud depth

Maximum diameter 60% of stud depth
Studs must be doubled
No more than two successive doubled studs so bored

$^5/_8$ in. minimum
from edge of stud

FIGURE 6-10 Boring and notching of studs in exterior wall or bearing interior wall

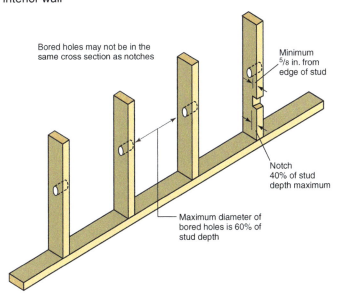

Bored holes may not be in the
same cross section as notches

Minimum
$^5/_8$ in. from
edge of stud

Maximum diameter of
bored holes is 60% of
stud depth

Notch
40% of stud
depth maximum

FIGURE 6-11 Boring and notching of studs in nonbearing interior wall

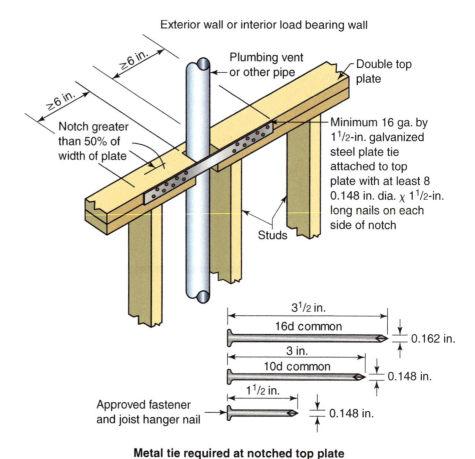

Exterior wall or interior load bearing wall

≥6 in.

≥6 in.

Plumbing vent or other pipe

Double top plate

Notch greater than 50% of width of plate

Minimum 16 ga. by 1¹/₂-in. galvanized steel plate tie attached to top plate with at least 8 0.148 in. dia. χ 1¹/₂-in. long nails on each side of notch

Studs

3¹/₂ in.
16d common
0.162 in.

3 in.
10d common
0.148 in.

1¹/₂ in.
Approved fastener and joist hanger nail
0.148 in.

Metal tie required at notched top plate

FIGURE 6-12 Drilling and notching of top plate in exterior wall or bearing interior wall

FIREBLOCKING

To stop the spread of fire in concealed spaces of wood frame construction, fireblocking is required to form an effective barrier between stories and between the top story and the attic. Concealed spaces of stud walls and partitions require fireblocking vertically at the ceiling and floor levels. In platform framing, the top wall plates typically satisfy the fireblocking requirement. The studs generally provide effective fireblocking in the horizontal direction, but for walls with offset studs or other openings, fireblocking is required horizontally at 10-foot intervals or less. Fireblocking also is required at all interconnections between concealed vertical and horizontal spaces, such as those created by soffits, and at the top and bottom of stair stringers (Figure 6-13). Openings around vents, pipes, ducts, cables and wires must also be sealed at the ceiling and floor level (Figure 6-14). Fireblocking materials include nominal 2-inch-thick lumber, equivalent layers of structural wood panels, and glass fiber insulation securely retained in place. **[Ref. R302.11]**

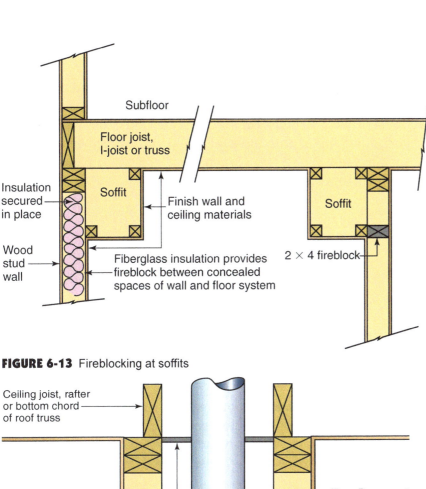

FIGURE 6-13 Fireblocking at soffits

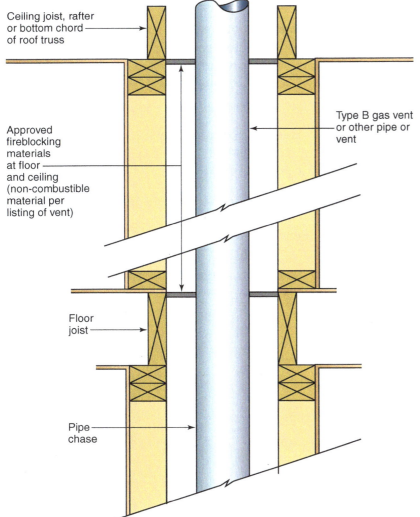

FIGURE 6-14 Fireblocking at pipe chase

DRAFTSTOPPING

When a ceiling is applied to the bottom side of open web floor trusses, or large areas of communicating spaces are otherwise created in the floor assembly, the IRC requires draftstopping to divide the horizontal spaces into areas of 1,000 square feet or less. One-half-inch gypsum board and ⅜-inch wood structural panels are approved draftstopping materials. [Ref. R302.12]

FLOORS

The conventional wood framing floor system includes the support beams or girders, floor joists, floor sheathing and connections necessary for the load path to the foundation. The IRC provides the lumber framing and wood structural panel spans in prescriptive tables. Floor framing details are shown in Figures 6-15 through 6-17. [Ref. R502]

Beams and girders

The prescriptive spans and support requirements for headers, beams, and girders are based on common species of #2 grade lumber and incorporate the required strength and deflection criteria under code-prescribed uniform loads. [Ref. Tables R6027(1), R602.7(2)]

EXAMPLE 6-1

Determine the minimum size and bearing support requirements for an interior beam of #2 hem-fir lumber supporting two floors. The width of the building is 24 feet, and the beam span is 6 feet. Refer to Table 6-2 and Figure 6-15.

TABLE 6-2 Maximum girder and header spans for interior bearing walls for # 2 Douglas fir-larch, hem-fir, southern pine and spruce-pine-fir, and required number of jack studs

| Headers and girders supporting | Size | Building width (feet) | | | | | |
| | | 12 | | 24 | | 26 | |
		Span	Jack studs	Span	Jack studs	Span	Jack studs
Two floors	2-2 × 4	2-7	1	1-11	1	1-7	1
	2-2 × 6	3-11	1	2-11	2	2-5	2
	2-2 × 8	5-0	1	3-8	2	3-1	2
	2-2 × 10	5-11	2	4-4	2	3-7	2
	2-2 × 12	6-11	2	5-2	2	4-3	3
	3-2 × 8	6-3	1	4-7	2	3-10	2
	3-2 × 10	7-5	1	5-6	2	4-6	2
	3-2 × 12	8-8	2	6-5	2	5-4	2
	4-2 × 8	7-2	1	5-4	1	4-5	2
	4-2 × 10	8-6	1	6-4	2	5-3	2
	4-2 × 12	10-1	1	7-5	2	6-2	2

[Ref. excerpt of Table R602.7(2)]

Spans are given in feet and inches.

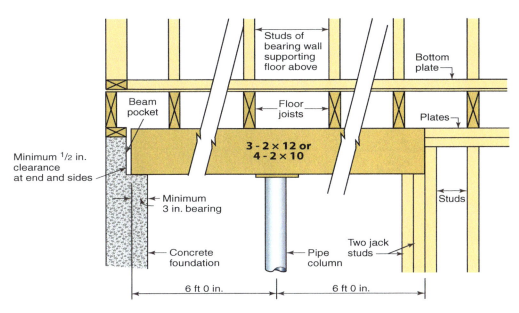

FIGURE 6-15 Solution for interior beam span and bearing support based on Table 6-2

Joists

In determining the minimum size of a floor joist for a given span, the designer or builder must consider a number of criteria, including live load, dead load, spacing of joists and the species and grade of lumber. The minimum live load for residential areas is 40 psf, except for sleeping areas, habitable attics and attics served by fixed stairs, where a live load of 30 psf is permitted. The assumed value for uniform dead load in conventional wood frame construction is typically 10 psf. The use of heavier flooring materials such as lightweight concrete or natural stone floor covering may necessitate a joist size based on 20 psf dead load. Table 6-3 contains typical fastening requirements for floor systems. [Ref. Tables R502.3.1(1) and R502.3.1(2)]

EXAMPLE 6-2
Determine the minimum size and maximum spacing of a floor joist of #2 Douglas fir-larch with a span of 14 feet. The floor joists are for a family room, dining room, and kitchen area, and the dead load is 10 psf. Refer to Table 6-3 and Figure 6-16.

TABLE 6-3 Floor joist spans for common lumber species, #2 grade (residential living areas, live load = 40 psf, L/360)

Joist spacing (Inches)	Species and grade		Dead load 5 10 psf			
			2 × 6	2 × 8	2 × 10	2 × 12
			Maximum floor joist spans			
			(ft - in.)	(ft - in.)	(ft - in.)	(ft - in.)
12	Douglas fir-larch	#2	10 - 9	14 - 2	18 - 0	20 - 11
	Hem-fir	#2	10 - 0	13 - 2	16 - 10	20 - 4
	Southern pine	#2	10 - 3	13 - 6	16 - 2	19 - 1
	Spruce-pine-fir	#2	10 - 3	13 - 6	17 - 3	20 - 7
16	Douglas fir-larch	#2	9 - 9	12 - 9	15 - 7	18 - 1
	Hem-fir	#2	9 - 1	12 - 0	15 - 2	17 - 7
	Southern pine	#2	9 - 4	11- 10	14 - 0	16 - 6
	Spruce-pine-fir	#2	9 - 4	12 - 3	15 - 5	17 - 10
19.2	Douglas fir-larch	#2	9 - 2	11 - 8	14 - 3	16 - 6
	Hem-fir	#2	8 - 7	11 - 3	13 - 10	16 - 1
	Southern pine	#2	8 - 6	10 - 10	12 - 10	15 - 1
	Spruce-pine-fir	#2	8 - 9	11 - 6	14 - 1	16 - 3
24	Douglas fir-larch	#2	8 - 3	10 - 5	12 - 9	14 - 9
	Hem-fir	#2	7 - 11	10 - 2	12 - 5	14 - 4
	Southern pine	#2	7 - 7	9 - 8	11 - 5	13 - 6
	Spruce-pine-fir	#2	8 - 1	10 - 3	12 - 7	14 - 7

[Ref. excerpt of Table R502.3.1(2)]

Spans are given in feet and inches.

Solution	
Joist spacing	Minimum joist size
12 in. oc	2 × 8
16 in. oc	2 × 10
19.2 in. oc	2 × 10
24 in. oc	2 × 12

Span: 14 ft 0 in.
Live load: 40 PSF
Dead load: 10 PSF
Lumber: No. 2 Douglas fir-larch

FIGURE 6-16 Floor joist span and bearing details based on Table 6-3

TABLE 6-4 Floor framing fastening schedule for typical box, sinker and pneumatic-driven nails

Description	Number and size of nails	Spacing, location and method
Joist to sill, top plate or girder	3 - 10d box (3" × 0.128") or 3 - 3" × 0.131" nails	Toe nail
Rim joist, band joist or blocking to sill or top plate	10d box (3" × 0.128") or 3" × 0.131" nails	6" o.c. toe nail
Band or rim joist to joist	4 - 10d box (3" × 0.128") or 4 - 3" × 0.131" nails	End nail
Built-up girders and beams, 2-inch lumber layers	10d box (3" × 0.128") or 3" × 0.131" nails and 3 - 10d box (3" × 0.128") or 3 - 3" × 0.131" nails	24" o.c. face nail at top and bottom staggered on opposite sides — Face nail at ends and at each splice

[Ref. excerpt of Table R602.3(1)]

Note: Some approved nail and staple sizes are not shown.

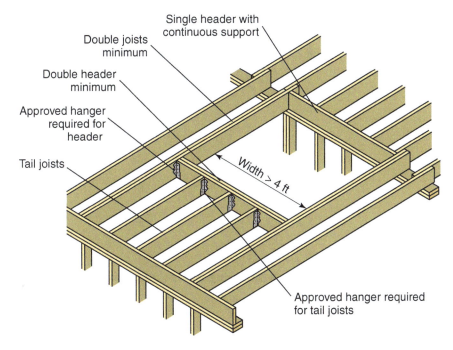

FIGURE 6-17 Framing at floor openings

DECKS

The IRC provides prescriptive methods for conventional wood deck construction that reflect widely accepted construction techniques in use over many decades. The provisions do not intend to limit design flexibility, and other approved methods may be used. Deck support provisions describe maximum joist and beam spans, appropriate joist spacing for the type of decking material, minimum post sizes, connections between beams and posts, and minimum bearing lengths. Details also are provided for attachment of the deck to the structure. Deck posts must be supported on footings and restrained at the bottom to prevent lateral movement. [Ref. R507]

Deck footings

The code prescribes the minimum size and depth of concrete deck footings based on the tributary area, snow or deck live load, and soil bearing pressure. The IRC Table provides prescriptive values for either square or round footings (Table 6-5 and Figure 6-18). See Figure 5-8 in Chapter 5 for more information on tributary loads. **[Ref. R507.3]**

EXAMPLE 6-3

Determine the minimum round concrete footing size for the deck corner post and interior post of a 20-foot × 12-foot free-standing deck based on Table 6-5 and Figure 6-18. The live load is 40 psf and exceeds the snow load. The presumed soil bearing pressure is 2000 psf.

TABLE 6-5   Minimum concrete footing size for decks

Live or Ground Snow Load (psf)	Tributary area (ft²)	Load Bearing Value of Soil (psf)					
		1500			2000		
		Side of a square footing (in)	Diameter of a round footing (in)	Thickness (in)	Side of a square footing (in)	Diameter of a round footing (in)	Thickness (in)
40	20	12	14	6	12	14	6
	40	14	16	6	12	14	6
	60	17	19	6	15	17	6
	80	20	22	7	17	19	6
	100	22	25	8	19	21	6
	120	24	27	9	21	23	7
	140	26	29	10	22	25	8
	160	28	31	11	24	27	9

[Ref. Excerpt of Table R507.3.1]

Spans are given in feet and inches.

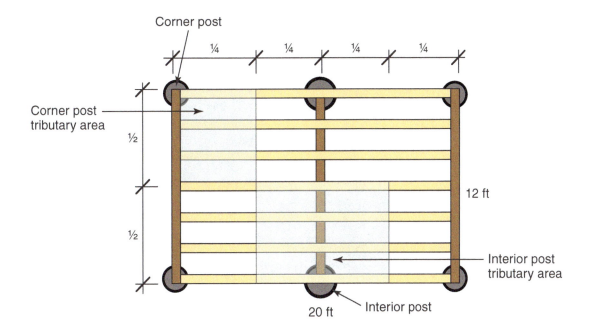

Tributary area – Corner post

Length is ¼ of total length = 20 ft × ¼ = 5 ft

Width is ½ of total width = 12 ft × ½ = 6 ft

Area = 5 ft × 6 ft = 30 ft²

Tributary area – Interior post

Length is ½ of total length = 20 ft × ½ = 10 ft

Width is ½ of total width = 12 ft × ½ = 6 ft

Area = 10 ft × 6 ft = 60 ft²

Footing size – Corner post

Min. 14 in. diameter

Min. 6 in. thick

Footing size – Interior post

Min. 17 in. diameter

Min. 6 in. thick

FIGURE 6-18 Determing concrete footing size based on deck tributary area and Table 6-5

Deck joists and beams

The joist and beam span values in Tables 6-6 and 6-7 assume outdoor, potentially wet conditions and include wood species such as redwood, western cedars, ponderosa pine and red pine that are commonly used in deck construction. Minimum bearing lengths match those for interior floor joists and beams: 1½ inches on wood or metal and 3 inches on concrete or masonry.

Joists framing into the side of a ledger board or beam require joist hangers (Figure 6-19). **[Ref. R507.5, R507.6, Table R507.5, Table R507.6]**

TABLE 6-6 Deck joist spans (feet – inches)

Species	Size	Allowable joist span			Maximum cantilever		
		Spacing of deck joists with no cantilever (in.)			Spacing of deck joists with cantilevers (in.)		
		12	**16**	**24**	**12**	**16**	**24**
Southern pine	2 × 6	9-11	9-0	7-7	1-3	1-4	1-6
	2 × 8	13-1	11-10	9-8	2-1	2-3	2-5
	2 × 10	16-2	14-0	11-5	3-4	3-6	2-10
	2 × 12	18-0	16-6	13-6	4-6	4-2	3-4
Douglas fir-larch, hem-fir, spruce-pine-fir	2 × 6	9-6	8-8	7-2	1-2	1-3	1-5
	2 × 8	12-6	11-1	9-1	1-11	2-1	2-3
	2 × 10	15-8	13-7	11-1	3-1	3-5	2-9
	2 × 12	18-0	15-9	12-10	4-6	3-11	3-3
Redwood, western cedars, ponderosa pine, red pine	2 × 6	8-10	8-0	7-0	1-0	1-1	1-2
	2 × 8	11-8	10-7	8-8	1-8	1-10	2-0
	2 × 10	14-11	13-0	10-7	2-8	2-10	2-8
	2 × 12	17-5	15-1	12-4	3-10	3-9	3-1

[Ref. excerpt of Table R507.6]

Note: Spans based on No. 2 grade lumber with wet service factor.

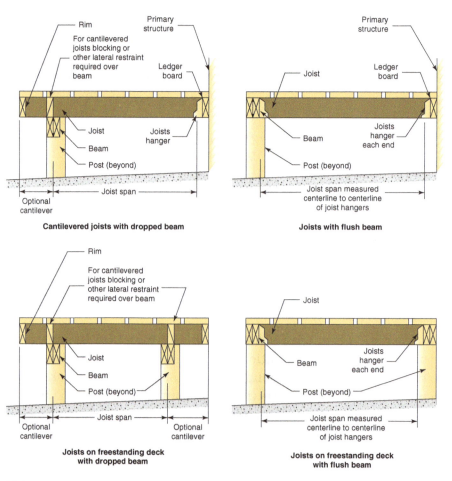

FIGURE 6-19 Typical deck joist span and supports

TABLE 6-7 Deck beam spans (feet - inches)

Species	Size	Deck joist span (ft.)				
		8	10	12	14	16
Southern pine	2 – 2 x 8	7-7	6-9	6-2	5-9	5-4
	2 – 2 x 10	9-0	8-0	7-4	6-9	6-4
	2 – 2 x 12	10-7	9-5	8-7	8-0	7-6
	3 – 2 x 8	9-6	8-6	7-9	7-2	6-8
	3 – 2 x 10	11-3	10-0	9-2	8-6	7-11
	3 – 2 x 12	13-3	11-10	10-9	10-0	9-4
Douglas fir-larch, hem-fir, spruce-pine-fir, redwood, western cedars, ponderosa pine, red pine	2 – 2 x 8	5-11	5-4	4-10	4-6	4-1
	2 – 2 x 10	7-3	6-6	5-11	5-6	5-1
	2 – 2 x 12	8-5	7-6	6-10	6-4	5-11
	3 – 2 x 8	8-6	7-7	6-11	6-5	6-0
	3 – 2 x 10	10-5	9-4	8-6	7-10	7-4
	3 – 2 x 12	12-1	10-9	9-10	9-1	8-6
[Ref. excerpt of Table R507.5]						

Note: Spans based on No. 2 grade lumber with wet service factor

Deck posts

Provisions for sizing wood deck posts are limited to single-level decks and are based on the height of the post measured to the bottom of the beam. The code prescribes either a notched post with two through-bolts or a manufactured post cap for connecting the post to the beam. In the case of a notched post, the post must be at least a 4 × 6 or 6 × 6 to provide a minimum cross section of 5½ inches for notching (Table 6-8, Figure 6-20). A manufactured connector is required for the post connection to the footing unless the post is embedded at least 12 inches in the ground or concrete pier. **[Ref. R507.4, R507.5, Table R507.4]**

TABLE 6-8 Deck post size and height for single-level wood-framed decks

Deck post size	Maximum height*(ft.)	Notes
4 x 4	8	When supporting one-ply or two-ply beams
4 x 4	6-9	When supporting three-ply beams on a post cap
4 x 6	8	
6 x 6	14	
8 x 8	14	
[Ref. Table R507.4]		

*Measured to the underside of the beam

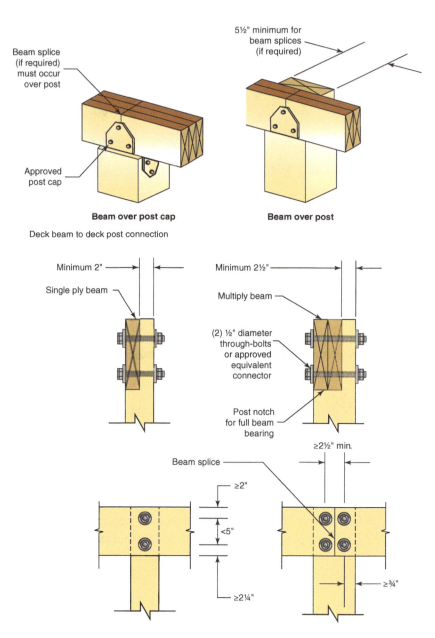

FIGURE 6-20 Connection of deck posts to deck beam

Deck attachment

The prescriptive methods for deck attachment apply to a minimum 2 by 8 deck ledger connected to a 2-inch nominal solid-sawn lumber band joist or a minimum 1-inch by 9½-inch Douglas fir laminated veneer lumber (LVL) rim board. Attachment to other structural composite lumber band joists requires a design in accordance with accepted engineering practice. Fasteners must be minimum ½-inch diameter hot-dipped galvanized or stainless steel lag screws or bolts installed with washers of the same material. The maximum spacing is based on the deck joist span. The code requires a staggered fastener pattern with the bolts or lag screws located not less than 2 inches from the top edge and not less than $3/4$ inch from the bottom edge of the deck ledger and from 2 to 5 inches from the end of the ledger (Table 6-9 and Figure 6-21). Other methods may still be used, and often are, to provide equivalent connection capacities, as long as the method is approved by the building official. For example, proprietary fasteners are commonly installed following the manufacturer's instructions and based on equivalent capacities. **[Ref. R507.9]**

TABLE 6-9 Fastener spacing for a southern pine or hem-fir deck ledger and a 2-inch nominal solid-sawn spruce-pine-fir band joist (deck live load = 40 psf, deck dead load = 10 psf)

Joist span	6'-0" and less	6'-1" to 8'-0"	8'-1" to 10'-0"	10'-1" to 12'-0"	12'-1" to 14'-0"	14'-1" to 16'-0"	16'-1" to 18'-0"
Connection details	On-center spacing of fasteners						
$1/2$" diameter lag screw with $1/2$" maximum WSP sheathing	30	23	18	15	13	11	10
$1/2$" diameter bolt with $1/2$" maximum WSP sheathing	36	36	34	29	24	21	19
$1/2$" diameter bolt with 1" maximum sheathing	36	36	29	24	21	18	16

[Ref. Table R507.9.1.3(1)]

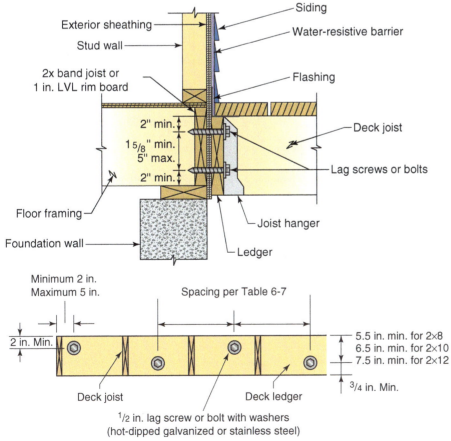

FIGURE 6-21 Deck ledger connection to band joist

WALLS

Walls must be designed and constructed to safely support all code-prescribed loads and transfer those loads to the supporting structure and foundation. In addition to setting limits on the size, length, and spacing of studs, the code includes a number of methods for wall bracing, which is critical to the structural integrity of the building. Wall framing details are shown in Figure 6-22. Table 6-10 contains typical fastening requirements for wall framing. [Ref. R602]

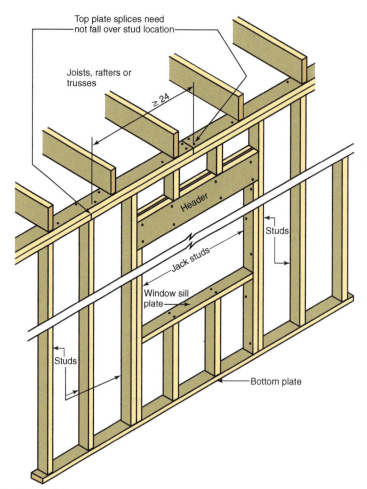

Top plate splices need not fall over stud location

Joists, rafters or trusses

≥ 24

Header

Studs

Jack studs

Window sill plate

Studs

Bottom plate

FIGURE 6-22 Typical wall framing details

TABLE 6-10 Wall framing fastening schedule for typical box, sinker and pneumatic-driven nails

Description	Number and size of nails	Spacing, location and method
Stud to stud (not at braced wall panels)	10d box (3" × 0.128") or 3" × 0.131" nails	16" o.c. face nail
Stud to stud and abutting studs at intersecting wall corners (at braced wall panels)	16d box (3½" × 0.135") or 3" × 0.131" nails	12" o.c. face nail
Built-up header (2" to 2" header with ½" spacer)	16d box (3½" × 0.135")	12" o.c. each edge face nail
Continuous header to stud	4 - 10d box (3" × 0.128")	Toe nail
Top plate to top plate	10d box (3" × 0.128") or 3" × 0.131" nails	12" o.c. face nail
Double top plate splice	12 - 10d box (3" × 0.128") or 12 - 3" × 0.131" nails	Face nail on each side of end joint (minimum 24" lap)
Bottom plate to joist, rim joist, band joist or blocking (not at braced wall panels)	16d box (3½" × 0.135") or 3" × 0.131" nails	12" o.c. face nail
Bottom plate to joist, rim joist, band joist or blocking (at braced wall panel)	3 - 16d box (3½" × 0.135") or 4 - 3" × 0.131" nails	3 each 16" o.c. face nail 4 each 16" o.c. face nail
Top or bottom plate to stud	3 - 10d box (3" × 0.128") or 3 - 3" × 0.131" nails	End nail
Top plates, laps at corners and intersections	3 - 10d box (3" × 0.128") or 3 - 3" × 0.131" nails	Face nail

[Ref. excerpt from Table R602.3(1)]

Note: Some approved nail and staple sizes are not shown.

Studs and plates

The prescriptive provisions of the IRC generally limit stud height in bearing walls to 10 feet. The height is the distance between points of lateral support perpendicular to the plane of the wall, which are typically where the top and bottom plates are connected to the floor or ceiling framing. The size and spacing of studs is related to the number of floors being supported with or without the additional load of the roof-ceiling assembly (Figure 6-23). [Ref. Table R602.3(5)]

Headers

Headers are required above door and window openings to carry the loads of construction above and transfer the loads to the wall framing at the sides of the opening. The prescriptive tables for girders and headers provide the span and bearing support requirements for dimension lumber headers. In addition, the IRC prescribes the minimum number of full-height (king) studs at each end of a header in an exterior wall to resist out-of-plane wind loads on the window or door opening. [Ref. Tables R602.7(1), R602.7(2) and R602.7.5]

EXAMPLE 6-3

Determine the minimum size, maximum height, and maximum spacing of standard studs in an exterior bearing wall, as shown in Figure 6-23. Refer to Table 6-11.

TABLE 6-11 Size, height and maximum spacing of wood studs

		Bearing walls				Nonbearing walls	
Stud size (inches)	Laterally unsupported stud height (feet)	Supporting a roof-ceiling assembly only (inches)	Supporting one floor, plus a roof-ceiling assembly (inches)	Supporting two floors, plus a roof-ceiling assembly (inches)	Supporting one floor only (inches)	Laterally unsupported stud height (feet)	Maximum spacing (inches)
2 × 3	—	—	—	—	—	10	16
2 × 4	10	24	16	—	24	14	24
3 × 4	10	24	24	16	24	14	24
2 × 5	10	24	24	—	24	16	24
2 × 6	10	24	24	16	24	20	24

[Ref. excerpt of Table R602.3(5)]

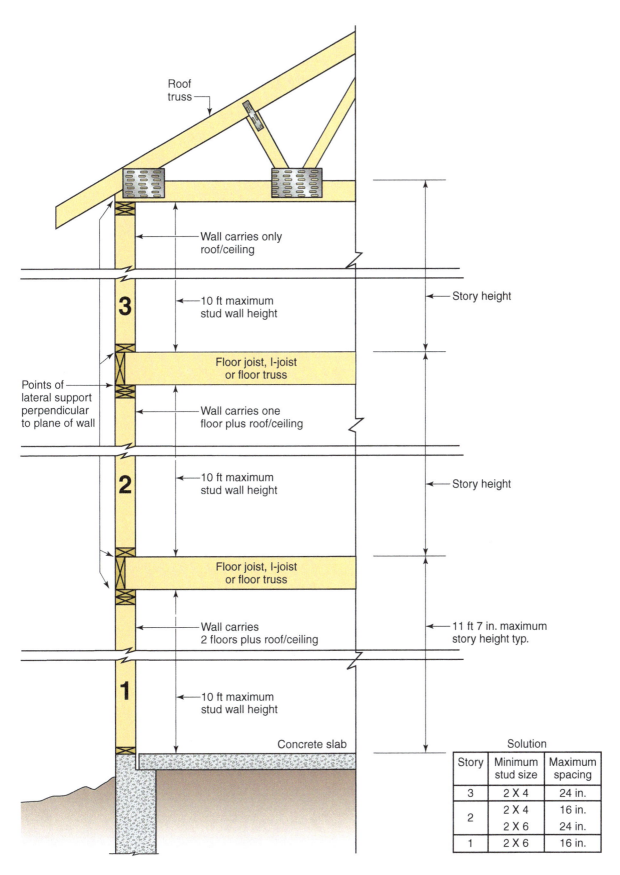

Roof truss

Wall carries only roof/ceiling

3   10 ft maximum stud wall height

Story height

Floor joist, I-joist or floor truss

Points of lateral support perpendicular to plane of wall

Wall carries one floor plus roof/ceiling

2   10 ft maximum stud wall height

Story height

Floor joist, I-joist or floor truss

Wall carries 2 floors plus roof/ceiling

11 ft 7 in. maximum story height typ.

1   10 ft maximum stud wall height

Concrete slab

Solution		
Story	Minimum stud size	Maximum spacing
3	2 X 4	24 in.
2	2 X 4	16 in.
	2 X 6	24 in.
1	2 X 6	16 in.

FIGURE 6-23 Stud size, height and spacing based on Table 6-11

EXAMPLE 6-4

Determine the minimum size and bearing support requirements for a #2 Douglas fir-larch header in an exterior bearing wall as shown in Figure 6-24. Determine the minimum number of full-height studs at each end of all headers. The width of the building is 24 feet, the header span is 7 feet and the ground snow load is 30 psf. Wind speed is 115 mph in wind exposure category C. Refer to Tables 6-12 and 6-13 and Figure 6-24.

TABLE 6-12 Girder spans and header spans for exterior bearing walls (maximum spans for # 2 grade Douglas fir-larch, hem-fir, southern pine, and spruce-pine-fir) and required number of jack studs

| Girders and headers supporting | Size | Ground snow load (psf) 30 — Building width (feet) | | | |
| | | 24 | | 36 | |
		Span	Jack studs	Span	Jack studs
Roof and ceiling	2 - 2 × 8	5 - 9	1	4 - 10	2
	2 - 2 × 10	6 - 10	2	5 - 9	2
Header Typ	2 - 2 × 12	8 - 1	2	6 - 10	2
	3 - 2 × 8	7 - 3	1	6 - 1	1
	3 - 2 × 10	8 - 7	1	7 - 3	2
	3 - 2 × 12	10 - 1	2	8 - 6	2
Roof, ceiling and one center-bearing floor	2 - 2 × 8	4 - 10	2	4 - 1	2
	2 - 2 × 10	5 - 8	2	4 - 10	2
	2 - 2 × 12	6 - 8	2	5 - 8	2
	3 - 2 × 8	6 - 0	1	5 - 1	2
	3 - 2 × 10	7 - 2	2	6 - 1	2
	3 - 2 × 12	8 - 5	2	7 - 2	2
Roof, ceiling and two center-bearing floors	2 - 2 × 8	4 - 0	2	3 - 5	2
	2 - 2 × 10	4 - 9	2	4 - 0	2
	2 - 2 × 12	5 - 7	2	4 - 9	3
	3 - 2 × 8	5 - 0	2	4 - 3	2
	3 - 2 × 10	5 - 11	2	5 - 1	2
	3 - 2 × 12	7 - 0	2	5 - 11	2

[Ref. excerpt of Table R602.7(1)]

Spans are given in feet and inches.

TABLE 6-13 Minimum number of full-height studs at each end of headers in exterior walls

Maximum header span (feet)	Ultimate design wind speed and exposure category	
	≤ 115 mph, Exposure B	< 140 mph, Exposure B or < 130 mph, Exposure C
4	1	1
6	1	2
8	1	2
10	2	3
12	2	3
14	2	3
16	2	4
18	2	4

Ref. Table R602.7.5

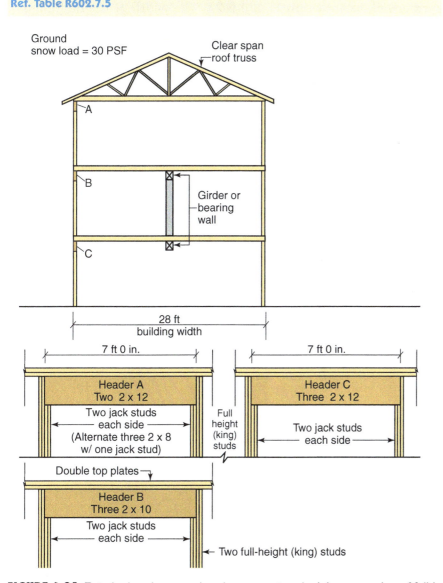

FIGURE 6-24 Exterior header span, bearing support and minimum number of full-height studs based on Tables 6-12 and 6-13

Single member headers

Traditionally, headers in conventional wood frame construction have consisted of at least two dimension lumber members side-by-side, each having a nominal dimension of 2 inches in thickness. As an alternative, the IRC allows single member headers of 2-inch nominal thickness under limited loading conditions to increase the energy efficiency of the dwelling and reduce the cost of construction. Installation of a single header results in a greater thickness of cavity insulation to reduce heat loss through the header in exterior walls. The prescriptive provisions for single headers are limited to areas with a maximum 50-psf ground snow load. **[Ref. R602.7.1, Tables R602.7(1)and R602.7.5]**

EXAMPLE 6-5

Determine the minimum size and the minimum number of jack studs and full-height studs for a single header in an exterior bearing wall, as shown in Figure 6-25. The width of the building is 24 feet, the header span is 4 feet 11 inches, and the ground snow load is 30 psf. Wind speed is 120 mph in wind exposure category B. Refer to Tables 6-13 and 6-14 and Figure 6-25.

TABLE 6-14 Single header spans for exterior bearing walls (maximum spans for # 2 grade Douglas fir-larch, hem-fir, southern pine, and spruce-pine-fir) and required number of jack studs

Single headers supporting	Size	Ground snow load (psf)	
		30	
		Building width (feet)	
		24	
		Span	Jack studs
Roof and ceiling	2 × 10	4–8	2
	2 × 12	5–5	2

[Ref. excerpt from Table R602.7(1)]

Spans are given in feet and inches.

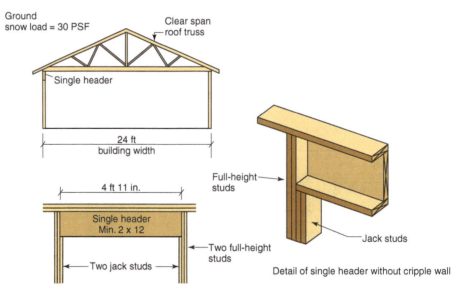

FIGURE 6-25 Single header in exterior bearing wall based on Tables 6-13 and 6-14

Wall bracing

Wall bracing is necessary to provide resistance to racking from lateral loads, primarily wind and seismic forces. The IRC includes twelve distinct, prescriptive methods of panel and diagonal wall bracing, referred to as intermittent bracing methods and four methods of continuous sheathing. Discussion in this chapter focuses on the most common bracing material—wood structural panels (Method WSP). A braced wall panel is the segment of bracing that is the full height of the wall, and its horizontal dimension is referred to as the length of the panel. The minimum length of a braced wall panel is typically 4 feet, but the code offers a number of alternatives to reduce this length by increasing the strength of the panel through specific material, connection, and anchorage details (Figures 6-26, 6-27, 6-28).

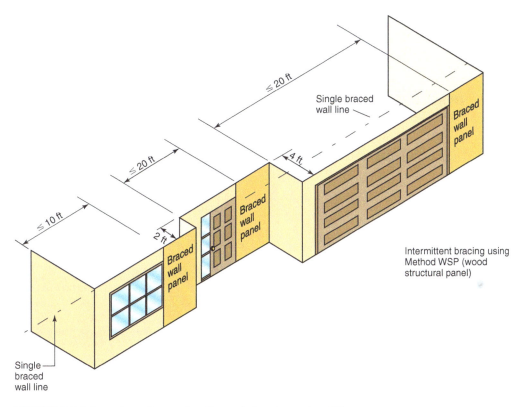

FIGURE 6-26 Location of braced wall panels in a braced wall line

You Should Know

Wood wall bracing

- Prescriptive methods
- Engineering not required but still an option
- Consists of sections of wall with sheathing nailed in a prescribed pattern
- Establishes minimum length of bracing along exterior walls and some interior walls
- Stiffens walls to resist forces pressing on exterior of building ●

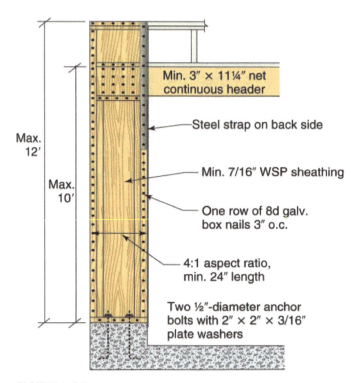

FIGURE 6-27 Method PFG—portal frame at garage door openings in Seismic Design Categories A, B and C

Min. 3″ × 11¼″ net continuous header

Steel strap on back side

Min. 7/16″ WSP sheathing

One row of 8d galv. box nails 3″ o.c.

4:1 aspect ratio, min. 24″ length

Two ½″-diameter anchor bolts with 2″ × 2″ × 3/16″ plate washers

Max. 12′

Max. 10′

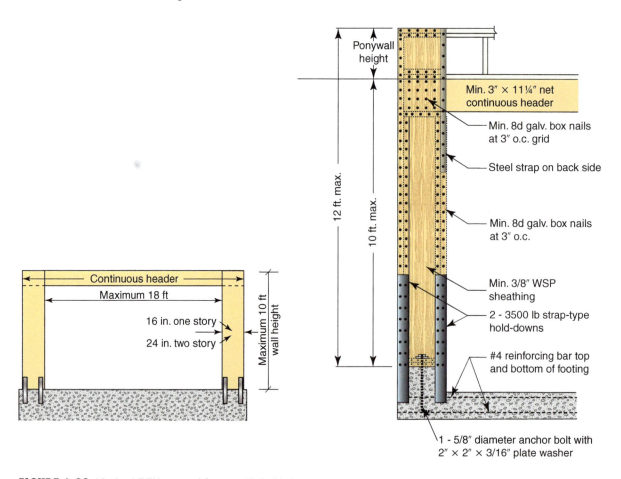

Ponywall height

Min. 3″ × 11¼″ net continuous header

Min. 8d galv. box nails at 3″ o.c. grid

Steel strap on back side

Min. 8d galv. box nails at 3″ o.c.

Min. 3/8″ WSP sheathing

2 - 3500 lb strap-type hold-downs

#4 reinforcing bar top and bottom of footing

1 - 5/8″ diameter anchor bolt with 2″ × 2″ × 3/16″ plate washer

12 ft. max.

10 ft. max.

Continuous header

Maximum 18 ft

16 in. one story

24 in. two story

Maximum 10 ft wall height

FIGURE 6-28 Method PFH—portal frame with hold-downs

The series of braced wall panels, typically in an exterior wall, is considered a braced wall line. A braced wall line must contain the prescribed total length of bracing in feet and meet the maximum spacing requirements of braced wall panels as well as braced wall lines. For this reason, interior braced wall lines also may be required. The amount and location of bracing is determined by numerous factors, including the number of stories of the building, the seismic design category, the design wind speed, the wind exposure category and the method of bracing. Another path for compliance with the wall bracing provisions is to apply structural panels to all areas of one side of a braced wall line, including above and below windows. An alternative to intermittent bracing, this continuous sheathing method increases the rigidity of the lateral resistance system and allows reduced lengths for full height braced wall panels. [Ref. R602.10]

CEILING AND ROOF

The scope of wood roof framing details in the code is limited to roofs with a minimum slope of 3 units horizontal to 12 units vertical (3:12).

Ceiling joist

In addition to supporting ceiling materials, ceiling joists also serve as rafter ties to resist the outward thrust of the rafters at the top of the wall. It follows that the ceiling joist requires adequate connection to the rafter, which is in turn fastened to the top of the wall. Maximum ceiling joist spans are provided for attics without storage and attics with limited storage. Attics with fixed stair access require joists sized as floor joists. [Ref. R802.5]

Rafters

The prescriptive tables giving maximum spans for rafters are based on the snow load of the geographic area (roof live load of 20 psf is used in areas with snow loads less than 30 psf) and whether the ceiling material is attached to the bottom of the rafter rather than a ceiling joist. Rafters line up opposite each other at the ridge and are typically framed to a ridge board, but gusset plate ties also are permitted. [Ref. R802.5]

Where ceiling joists are not connected to the rafters at the top plate or are installed perpendicular to the rafters, minimum 2 × 4 rafter ties are required to resist the outward thrust forces of the rafters on the wall. In the absence of joists or rafter ties, the ridge must be supported by a bearing wall or girder, or be designed as a beam (Figures 6-29 through 6-31). [Ref. R802.3, R802.4]

TABLE 6-15 Rafter spans for common lumber species, # 2 grade (ground snow load = 30 psf, ceiling not attached to rafters, L/180, dead load = 10 psf)

Rafter spacing (inches)	Species and grade		Dead load = 10 psf				
			2×4	2×6	2×8	2×10	2×12
			Maximum rafter spans				
			ft - in.	ft - in.	ft - in.	ft - in.	ft - in.
16	Douglas fir-larch	#2	8-3	12-1	15-4	18-9	21-8
	Hem-fir	#2	8-0	11-9	14-11	18-2	21-1
	Southern pine	#2	7-6	11-2	14-2	16-10	19-10
	Spruce-pine-fir	#2	8-2	11-11	15-1	18-5	21-5
24	Douglas fir-larch	#2	6-9	9-10	12-6	15-3	17-9
	Hem-fir	#2	6-7	9-7	12-2	14-10	17-3
	Southern pine	#2	6-1	9-2	11-7	13-9	16-2
	Spruce-pine-fir	#2	6-8	9-9	12-4	15-1	17-6

[Ref. excerpt of Table R802.4.1(3)]

Spans are given in feet and inches.

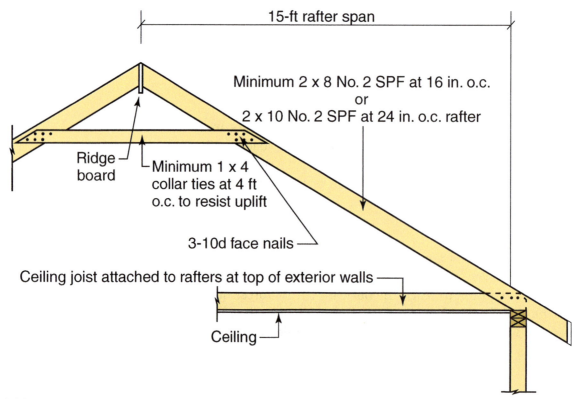

FIGURE 6-29 Rafter span with no ceiling attached based on Table 6-15

TABLE 6-16 Roof framing fastening schedule for typical box, sinker and pneumatic-driven nails

Description	Number and size of nails	Spacing, location and method
Blocking between ceiling joists or rafters to top plate	Three 10d box (3" × 0.128") or Three 3" × 0.131" nails	Toe nail
Ceiling joists to top plate	Three 10d box (3" × 0.128") or Three 3" × 0.131" nails	Per joist, toe nail
Ceiling joist not attached to parallel rafter, laps over partitions	Four 10d box (3" × 0.128") or Four 3" × 0.131" nails	Face nail
Ceiling joist attached to parallel rafter (heel joint)	Table R802.5.2	Face nail
Rafter or roof truss to plate	Three 16d box (3½" × 0.135") or Four 10d box (3" × 0.128") or Four 3" × 0.131" nails	2 toe nails on one side and 1 toe nail on opposite side of each rafter or truss
Roof rafters to ridge, valley or hip	Four 16d box (3½" × 0.135") or Four 10d box (3" × 0.128") or Four 3" × 0.131" nails	Toe nail
	Three 16d box (3½" × 0.135") or Three 10d box (3" × 0.128") or Three 3" × 0.131" nails	End nail

[Ref. excerpt from Table R602.3(1)]

Note: Some approved nail and staple sizes are not shown.

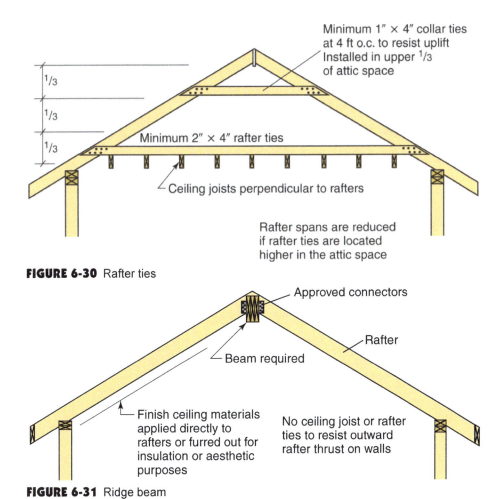

Minimum 1″ × 4″ collar ties at 4 ft o.c. to resist uplift Installed in upper 1/3 of attic space

1/3

1/3

1/3

Minimum 2″ × 4″ rafter ties

Ceiling joists perpendicular to rafters

Rafter spans are reduced if rafter ties are located higher in the attic space

FIGURE 6-30 Rafter ties

Approved connectors

Rafter

Beam required

Finish ceiling materials applied directly to rafters or furred out for insulation or aesthetic purposes

No ceiling joist or rafter ties to resist outward rafter thrust on walls

FIGURE 6-31 Ridge beam

Roof uplift connections

Roof-to-wall connections to resist wind uplift forces are one component of a complete load path necessary to transmit all loads bearing on a structure through the other structural elements to the foundation. The IRC provides prescriptive values for roof uplift resistance based on building width, wind speed, exposure category, and roof pitch. In many cases, the fastener schedule for wood structural members provides for adequate connection through conventional nailing methods. In the case of rafters or roof trusses, this toe-nail connection is achieved with three 16d box nails or three 10d common nails. Two toe nails are placed on one side and one on the other side of the truss or rafter and driven into the top wall plate. However, when the table values for uplift exceed 200 pounds, or the building otherwise exceeds the specified wind speed, exposure or building width criteria, a manufactured connector is required. The connector must have a rated capacity meeting or exceeding the uplift value for connecting each rafter or truss to the wall (Table 6-17 and Figure 6-32). As an alternative to the table values in the IRC, the manufactured roof truss design drawings may be used for determining minimum uplift resistance. **[Ref. R802.11, Table R802.11]**

TABLE 6-17 Rafter or truss uplift connection forces from wind (pounds per connection)

Rafter or truss spacing	Roof span (feet)	Exposure B	
		Ultimate design wind speed (mph)	
		115	
		Roof pitch	
		< 5:12	≥ 5:12
16 in. o.c.	28	132	117
	32	145	129
	36	160	141
24 in. o.c.	28	198	176
	32	218	194
	36	240	212

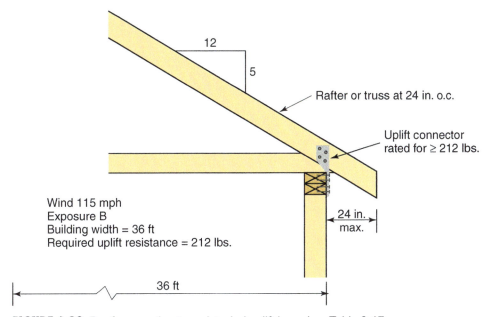

12

5

Rafter or truss at 24 in. o.c.

Uplift connector
rated for ≥ 212 lbs.

Wind 115 mph
Exposure B
Building width = 36 ft
Required uplift resistance = 212 lbs.

24 in.
max.

36 ft

FIGURE 6-32 Roof connection to resist wind uplift based on Table 6-17

Attic ventilation and access

The code requires cross ventilation for each attic or enclosed roof space to prevent moisture from accumulating in the space and causing damage to the structure. In poorly ventilated attics, warm moist air escaping from the conditioned space condenses on the framing and sheathing of the cooler attic space. The total net free ventilating area must be at least $1/150$ of the area of the space. A reduction to $1/300$ the area of the space is permitted when 40 to 50 percent of the required ventilating area is in the upper portion of the space, with the balance of the ventilating area provided by eave or cornice vents. This reduction is also permitted for attics in cold climates when a vapor retarder is installed on the warm-in-winter side of the ceiling. Unvented attics are permitted under certain conditions. [Ref. R806]

Access is required to attic areas with a ceiling height of 30 inches or more over an area of 30 square feet. The access requires a rough opening of at least 22 by 30 inches with headroom above the opening at least 30 inches high. The access opening must be located in a hallway or other readily accessible location. [Ref. R807]

IV

Finishes and Weather Protection

Chapter 7: Interior and Exterior Finishes and Weather Protection

Interior and Exterior Finishes and Weather Protection

As part of its purpose statement to protect the health and general welfare of the public, the *International Residential Code* (IRC) sets minimum requirements for durable interior and exterior finishes. Exterior wall and roof coverings protect the structure against the damaging effects of water intrusion and provide an envelope for a healthy and livable interior environment.

INTERIOR FINISHES

The IRC includes minimum installation requirements for gypsum board (drywall), plaster, ceramic tile and wood paneling for walls and ceilings. Inspection is not specifically required for other than attachment of lath or gypsum board that is part of a fire resistant rated assembly. Fire resistance requirements for walls and ceilings are covered in Chapter 9. Interior finishes require protection from moisture and are typically installed at a stage of construction when the building is substantially weather tight. The IRC does not regulate the installation of floor coverings or the application of paint and wallpaper. [Ref. R702]

Gypsum board

Gypsum board is the generic term for sheet panel products with a noncombustible gypsum core and paper facing that is commonly used for the wall and ceiling finish of dwelling construction. When taped and finished, it is often referred to as drywall. Type X gypsum board contains core additives for greater fire resistance than regular gypsum board. Gypsum products may also be reinforced for greater strength or manufactured for greater water resistance or durability.

The minimum fastening and thickness requirements for gypsum board relate to the spacing of the framing members and the location and intended application of the gypsum board. Generally, gypsum board may be installed with the long dimension perpendicular or parallel to framing members. The code requires installation perpendicular to ceiling framing members for ⅜-inch material and places additional limitations on ceiling applications receiving water-based textures or serving as a fire resistant rated separation. Tables 7-1 and 7-2 summarize gypsum board application requirements. [Ref. R702.3, Table R702.3.5]

Backing for ceramic tile and other nonabsorbent finishes

To prevent deterioration of the substrate and damage to the wall structure, only fiber-cement, fiber-mat-reinforced cementitious backer units, glass mat gypsum backers, or fiber-reinforced gypsum backers are permitted as backing material for wall tile installed in tub and shower areas. This requirement also applies to backers for nonabsorbent plastic panels in showers. In addition, installation of water-resistant gypsum backing board is not allowed where directly exposed to water or in areas subject to continuous high humidity. Water-resistant gypsum backing board is permitted on ceilings where framing spacing does not exceed 12 inches on center (O.C.) for ½-inch material or 16 inches O.C. for ⅝-inch material. [Ref. R702.3.7, R702.4.2]

TABLE 7-1 Minimum thickness and application of gypsum board

Thickness of gypsum board (in.)	Orientation of gypsum board to framing	Maximum spacing of framing members (in. o.c.)	Maximum spacing of fasteners (in.)	
			Nails	Screws
Wall Application				
⅜	Either direction	16	8	16
½	Either direction	24	8	12
		16	8	16
⅝	Either direction	24	8	12
		16	8	16
Ceiling application, no water-based texture material				
⅜*	Perpendicular	16	7	12
½	Either direction	16	7	12
	Perpendicular	24	7	12
⅝	Either direction	16	7	12
	Perpendicular	24	7	12
Ceiling application with water-based texture material				
½	Perpendicular	16	7	12
½ sag-resistant	Perpendicular	24	7	12
⅝	Perpendicular	16	7	12
	Perpendicular	24	7	12
Garage ceiling application with habitable space above				
⅝ Type X	Perpendicular	16	6	6
	Perpendicular	24	6	6

[Ref. Table R702.3.5]

*Not permitted to support insulation. o.c. = on center.

TABLE 7-2 Fasteners for the application of gypsum board

Thickness of gypsum board (in.)	Screws		Nails				
	Attached to steel framing	Attached to wood framing	Attached to wood framing				
	Type S	Type W or Type S	13 gauge	Ring-shank	Cooler	Gypsum board nail	
		Minimum length (in.)					
⅜	¾	1	1 ¼	1 ¼	1 ⅜		
½	⅞	1 ⅛	1 ⅜	1 ¼	1 ⅝	1 ⅝	
⅝	1	1 ¼	1 ⅝	1 ⅜	1 ⅞	1 ⅞	
⅝ Type X at garage ceiling beneath habitable rooms	1	1 ¼	—	—	1 ⅞	1 ⅞	

[Ref. Table R702.3.5]

EXTERIOR WALL COVERINGS

Water-resistant barriers, flashing, windows, doors and siding or veneers form the protective exterior wall envelope of a dwelling. [Ref. R703]

Water and moisture management

In wood or steel light-frame construction, including detached accessory buildings, the code requires a water resistant barrier over the sheathing of all exterior walls. Siding and veneers are typically not impervious to wind-driven rain, and the water-resistive barrier in combination with flashings completes the weather protective system to keep moisture out of the wall assembly. The IRC prescribes one layer of No. 15 asphalt felt applied horizontally with 2-inch laps for the water-resistive barrier, but approved house wrap and other materials tested to perform equivalently to the felt satisfy the requirement. House wraps must be installed in accordance with the manufacturer's instructions to shed water away from the sheathing to the outside of the wall coverings. **[Ref. R703.1, R703.2]**

Flashing

To prevent water entering behind exterior wall coverings and penetrating the wall assembly, the code requires corrosion-resistant flashing at specific locations, including exterior window and door openings, penetrations, projections, wall and roof intersections, and intersections of dissimilar materials. Flashing is particularly important for the ledger attachment joining a porch or deck to the house structure and protects not only the concealed framing but the structural integrity of the deck or porch (Figure 7-1). **[Ref. R703.4]**

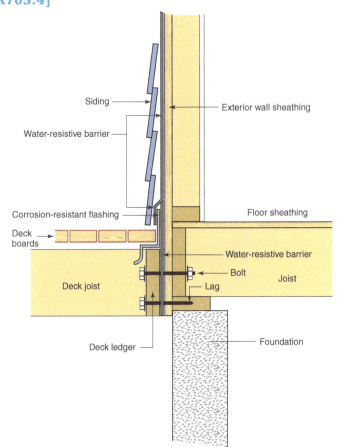

Siding
Water-resistive barrier
Corrosion-resistant flashing
Deck boards
Deck joist
Deck ledger

Exterior wall sheathing
Floor sheathing
Water-resistive barrier
Bolt
Joist
Lag
Foundation

FIGURE 7-1 Wall flashing at deck

Code Essentials

Water-resistive barrier for exterior plaster (stucco) and adhered masonry veneer

- Two layers of Grade D paper are required over sheathing.
- Grade D paper is a water-vapor permeable grade of Type 1 felt.
- Other house wrap products are permitted as alternates if they provide equivalent performance to two layers of Grade D paper.●

Code Essentials

Water-resistive barrier for exterior walls

- Felt paper or approved house wrap
- Required under all types of veneer and siding
- Covers all wall sheathing of dwelling including unheated areas
- Required for all buildings including detached accessory buildings
- Upper layer laps lower layer by 2 inches
- Laps over flashings to carry moisture to the outside
- Vertical laps of 6 inches ●

Masonry and stone veneer

Because of its vulnerability to damage due to earthquake forces, the prescriptive provisions of the IRC place greater limitations on the height, thickness, and weight of masonry and stone veneers in the higher seismic design categories (SDCs). For buildings sited in SDC A, B or C, the code generally permits veneers up to three stories and 30 feet above noncombustible foundations, with an additional 8 feet for gable end walls. Maximum thickness is 5 inches and the maximum weight is 50 psf. The code reduces the maximum height, thickness, and weight of veneer for buildings located in SDC D_0, D_1 or D_2. See Figure 7-2 for typical masonry veneer details. **[Ref. R703.8, Table R703.8(1), Table R703.8(2)]**

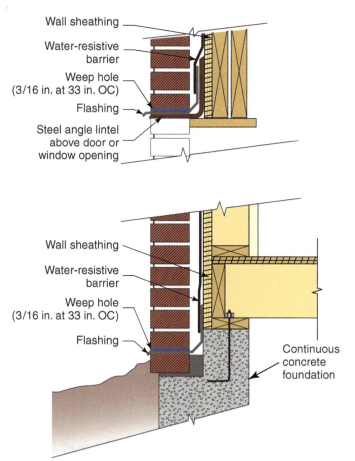

Wall sheathing

Water-resistive barrier

Weep hole (3/16 in. at 33 in. OC)

Flashing

Steel angle lintel above door or window opening

Wall sheathing

Water-resistive barrier

Weep hole (3/16 in. at 33 in. OC)

Flashing

Continuous concrete foundation

FIGURE 7-2 Brick veneer details

Support

Masonry veneer typically is supported by a continuous concrete or masonry foundation. In SDC A, B or C, light-frame construction may support exterior veneer weighing not more than 40 psf when designed to limit deflection to $1/_{600}$ of the span of the supporting members. Steel or noncombustible lintels are required above openings and must have bearing support of at least 4 inches at each end. Steel lintels require a rust-inhibitive shop coat on all surfaces or otherwise be protected against corrosion (Table 7-3, Figure 7-3). **[Ref. R703.8.2, R703.8.3]**

EXAMPLE 7-1
Based on Table 7-3, determine the minimum size of a steel lintel supporting masonry veneer with one story above. The width of the opening is 6 feet 0 inches. The solution is shown in Figure 7-3.

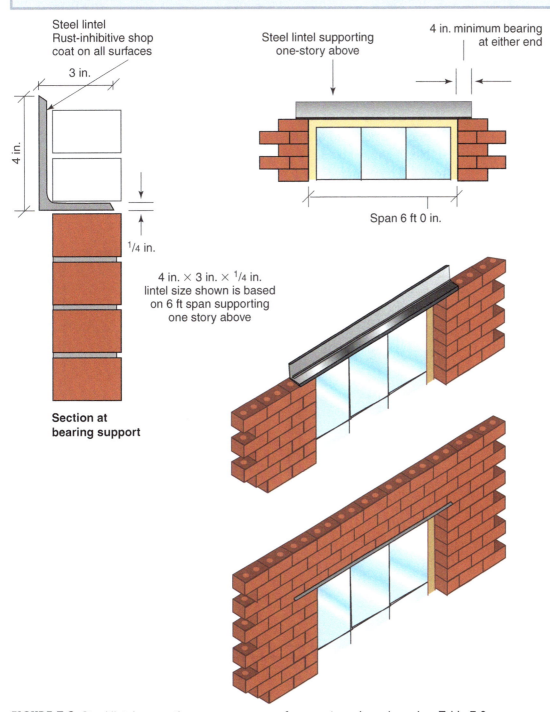

Steel lintel
Rust-inhibitive shop coat on all surfaces

3 in.

4 in.

1/4 in.

4 in. × 3 in. × 1/4 in. lintel size shown is based on 6 ft span supporting one story above

Section at bearing support

Steel lintel supporting one-story above

4 in. minimum bearing at either end

Span 6 ft 0 in.

FIGURE 7-3 Steel lintel supporting masonry veneer for one story above based on Table 7-3

TABLE 7-3 Allowable spans for steel lintels supporting masonry veneer

Size of steel angle (in.)	No story above	One story above	Two stories above
3 × 3 × ¼	6'-0"	4'-6"	3'-0"
4 × 3 × ¼	8'-0"	6'-0"	4'-6"
5 × 3½ × ⁵⁄₁₆	10'-0"	8'-0"	6'-0"
6 × 3½ × ⁵⁄₁₆	14'-0"	9'-6"	7'-0"
2 × 6 × 3½ × ⁵⁄₁₆	20'-0"	12'-0"	9'-6"

[Ref. Table R703.8.3.1]

Note: Long leg of the angle shall be placed in a vertical position.

Veneer anchoring

Veneer is anchored to the structure with corrosion-resistant metal ties of No. 9-gauge strand wire or No. 22-gauge × ⅞-inch corrugated sheet metal. The number of anchors is increased for installations in SDCs D_0, D_1 and D_2 and for townhouses in SDC C (Figures 7-4 and 7-5). **[Ref. R703.8.4, Table R703.8.4]**

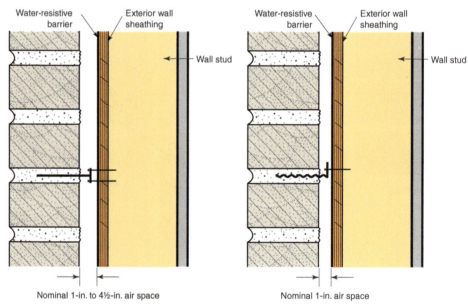

No. 9 gage wire tie with hook embedded in mortar joint

No. 22 gage x 7/8-in. wide corrugated steel tie

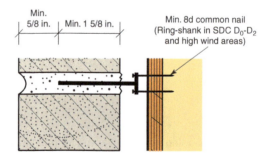

FIGURE 7-4 Brick veneer attachment and air space requirements

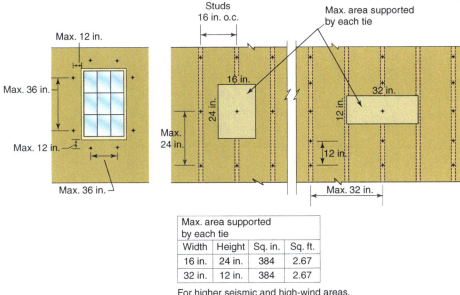

Max. area supported by each tie			
Width	Height	Sq. in.	Sq. ft.
16 in.	24 in.	384	2.67
32 in.	12 in.	384	2.67

For higher seismic and high-wind areas, each tie supports no more than 2.0 sq. ft.

FIGURE 7-5 Veneer tie spacing for all buildings in SDC A and B and detached dwellings in SDC C

Siding

Vertical panel siding and horizontal lap siding must be installed with proper joint treatments and flashing to resist penetration of moisture. Vertical joints require batten covers or a combination of flashing and sealants. Horizontal joints of panel siding require a minimum 1-inch lap, shiplap or a "Z" flashing over solid backing. Horizontal lap siding must be installed in accordance with the manufacturer's recommendations. In the absence of recommendations, the code requires a minimum lap of 1 inch to 1¼ inches depending on the siding material. All siding requires secure attachment with approved corrosion-resistant fasteners, which generally must penetrate into the wood framing 1 to 1½ inches, depending on the type of wall sheathing, siding material and the manufacturer's recommendations (Figure 7-6). Vinyl siding must be installed in accordance with the manufacturer's installation instructions and meet other applicable requirements based on design wind speed and wind exposure category (Figure 7-7). [Ref. R703.3 through R703.6, R703.10, R703.11, Table R703.3(1)]

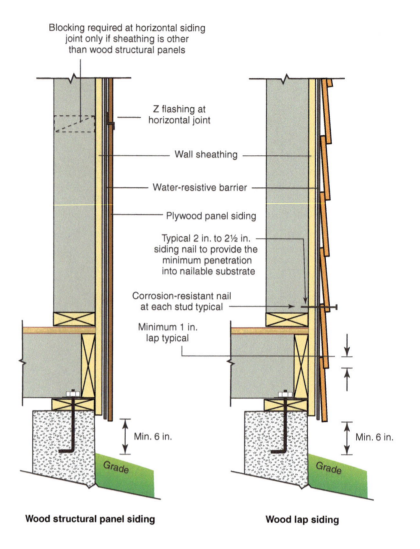

Blocking required at horizontal siding joint only if sheathing is other than wood structural panels

Z flashing at horizontal joint

Wall sheathing

Water-resistive barrier

Plywood panel siding

Typical 2 in. to 2½ in. siding nail to provide the minimum penetration into nailable substrate

Corrosion-resistant nail at each stud typical

Minimum 1 in. lap typical

Min. 6 in.

Grade

Min. 6 in.

Grade

Wood structural panel siding

Wood lap siding

FIGURE 7-6 Siding details

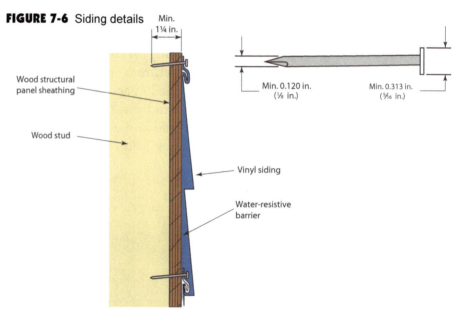

Min. 1¼ in.

Wood structural panel sheathing

Wood stud

Vinyl siding

Water-resistive barrier

Min. 0.120 in. (⅛ in.)

Min. 0.313 in. (⁵⁄₁₆ in.)

FIGURE 7-7 Vinyl siding attachment over wood structural panel sheathing

Exterior insulation finish system (EIFS)

EIFS, sometimes referred to as synthetic stucco due to its textured appearance, consists of a water-resistive barrier, drainage plane, rigid polystyrene insulation, a reinforced base coat and a trowel-applied textured finish coat. Where installed over frame construction, the code requires EIFS with drainage to provide a means to drain any moisture trapped behind the EIFS to the exterior. To protect the integrity of the weather repellant surface and prevent moisture penetration, face nailing of trim through the EIFS is not permitted. A minimum of 6 inches clearance is required between the ground and the lowest edge of the EIFS. In addition to the above code requirements, all EIFS must be installed in accordance with the manufacturer's installation instructions. [Ref. R703.9]

Windows

In the installation of windows and exterior doors (both known as fenestration), the code is concerned with the strength of the unit and its attachment to the structure to resist applicable wind loads, as well as the water resistance of the installation in the wall assembly. In this regard, fenestration must be manufactured to comply with the applicable referenced standards, and installation must follow the manufacturer's written instructions. The IRC also requires the manufacturer to provide installation and flashing detail instructions with each window and exterior door. Many manufacturers require a pan flashing installed at the base of the opening to direct water to the exterior. If installation instructions are not available, the code requires pan flashing (Figure 7-8). [Ref. R609, R703.4]

ROOF COVERING

The IRC prescribes the design, materials, construction and quality of roofing assemblies to provide weather protection for the building. Roof coverings must be installed according to the code and the manufacturer's instructions. This section will focus on the installation of asphalt and wood shingles, as well as the associated underlayment and flashing requirements.

Underlayment and ice barrier

At a minimum, for slopes of 4:12 or greater, one layer of No. 15 asphalt-saturated organic felt or other approved material is required to cover the roof deck before application of shingles. The code requires horizontal laps of at least 2 inches with end laps offset at least 6 feet in successive courses of felt (Figure 7-9).

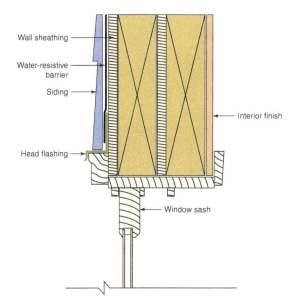
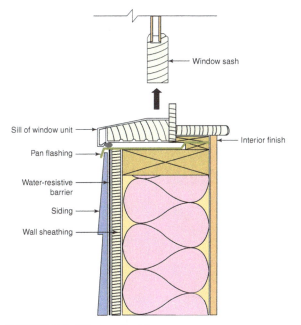

FIGURE 7-8 Window flashing

Slopes of at least 2:12 and less than 4:12 require two layers of felt with 19-inch horizontal overlaps (Figure 7-10). **[Ref. R905.1.1]**

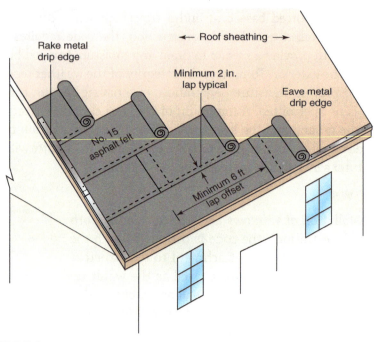

FIGURE 7-9 Underlayment for asphalt shingles on slopes of 4:12 or greater

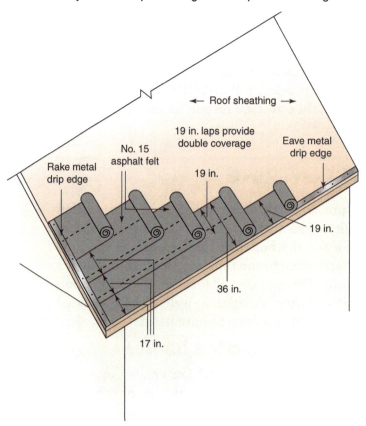

FIGURE 7-10 Underlayment for low slope application of asphalt shingles, slopes 2:12 or greater and less than 4:12

In cold climates, ice dams often form in gutters and along eaves. Freeze-thaw cycles combined with warm air escaping from the conditioned space thaws ice above the eave area. Water unable to drain past the ice dam is often forced back under the shingles and underlayment, causing damage to the structure beneath. In areas with a history of water damage to structures from ice dams at roof eaves, an ice barrier is required for added protection. The ice barrier consists of self-adhering polymer-modified bitumen material or two layers of cemented underlayment, which must extend from the eave to at least 24 inches inside the exterior wall line of the building (Figure 7-11). As discussed in Chapter 4, the jurisdiction indicates the ice barrier requirement in the "Climatic and Geographic Design Criteria" table when adopting the IRC. As damage from ice damming is less likely to occur, ice barriers are not required in unheated detached accessory buildings. [Ref. R905.1.2]

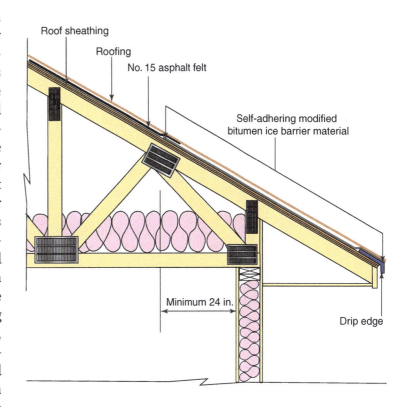

FIGURE 7-11 Ice barrier

Flashing

To effectively seal against entry of water, flashing is required at roof/wall intersections, at points of change in slope or direction, and around roof openings or penetrations. Where the roof joins a sidewall, either overlapping individual flashings at each course of shingles (Figure 7-12) or continuous flashing is required. The code requires flashing to be corrosion-resistant metal at least 0.019 inch thick (No. 26 galvanized sheet). Any chimney penetration more than 30 inches wide requires a cricket or saddle to divert water from the roof above to each side of the chimney. Crickets may be the same material as the roof covering or sheet metal (see Chapter 11). [Ref. R903.2, R905.2.8]

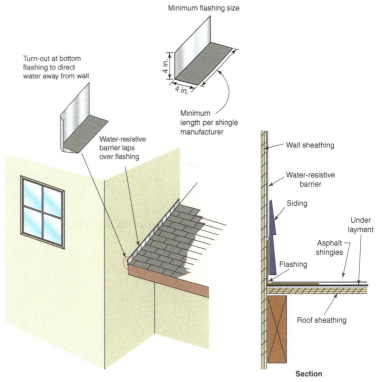

FIGURE 7-12 Sidewall flashing for asphalt shingles

Asphalt shingles

Asphalt shingles require a roof slope of at least 2:12 and must be installed in accordance with the manufacturer's instructions. Adequate wind resistance is addressed through testing to the appropriate standard for self-sealing or unsealed shingles and labeling to indicate the applicable classification for the design wind speed of the geographic location. Fasteners must be galvanized steel, stainless steel, aluminum, or copper roofing nails of at least 12 gauge (0.105 inch) with a head diameter not less than ⅜ inch. Nails must penetrate at least ¾ inch into the roof sheathing or penetrate through the sheathing. **[Ref. R905.2]**

The code recognizes a number of accepted practices for valley construction for asphalt shingles. Closed valleys (covered with shingles) require a valley lining consisting of a self-adhering polymer-modified bitumen sheet, an approved 24-inch-wide metal valley, one ply of approved smooth roll roofing at least 36 inches wide or two plies of mineral-surfaced roll roofing (Figure 7-13). Open valleys consist of an approved 24-inch-wide metal valley or two plies of mineral-surfaced roll roofing. **[Ref. R905.2.8.2]**

Wood shingles and wood shakes

The IRC includes requirements specific to wood shingles and shakes used for roofing. In addition to material and grading requirements, the code provides installation details related to slope, decking, underlayment, laps and exposure. Fastening must be in accordance with the manufacturer's instructions. The minimum slope is 3:12. Valleys require at least No. 26-gauge corrosion-resistant sheet metal extended 10 inches from the centerline each way for wood shingles and 11 inches from the centerline each way for shakes. End laps must be at least 4 inches (Tables 7-4 and 7-5 and Figures 7-14 and 7-15). **[Ref. R905.7, R905.8]**

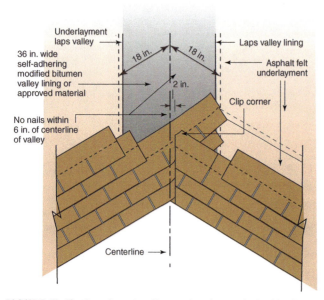

FIGURE 7-13 Cut closed valley option for asphalt shingles

TABLE 7-4 Wood shingle weather exposure and roof slope

Roofing material	Length (in.)	Grade	Exposure (in.)	
			3:12 pitch to < 4:12	4:12 pitch or steeper
Shingles of naturally durable wood	16	No. 1	3¾	5
		No. 2	3½	4
		No. 3	3	3½
	18	No. 1	4¼	5½
		No. 2	4	4½
		No. 3	3½	4
	24	No. 1	5¾	7½
		No. 2	5½	6½
		No. 3	5	5½

[Ref. Table R905.7.5(1)]

TABLE 7-5 Wood shake weather exposure and roof slope

Roofing material	Length (in.)	Grade	Exposure (in.), 4:12 pitch or steeper
Shakes of naturally durable wood	18	No. 1	7½
	24	No. 1	10*
Preservative-treated taper-sawn shakes of southern yellow pine	18	No. 1	7½
	24	No. 1	10
	18	No. 2	5½
	24	No. 2	7½
Taper-sawn shakes of naturally durable wood	18	No. 1	7½
	24	No. 1	10
	18	No. 2	5½
	24	No. 2	7½

[Ref. Table R905.8.6]

*For 24-in. by ⅜-in. handsplit shakes, the maximum exposure is 7½ in.

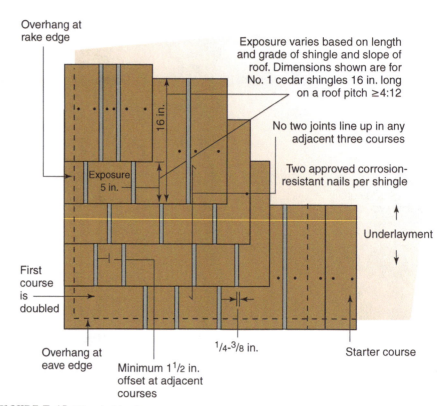

Overhang at rake edge

Exposure varies based on length and grade of shingle and slope of roof. Dimensions shown are for No. 1 cedar shingles 16 in. long on a roof pitch ≥4:12

16 in.

No two joints line up in any adjacent three courses

Two approved corrosion-resistant nails per shingle

Exposure 5 in.

Underlayment

First course is doubled

Overhang at eave edge

Minimum 1½ in. offset at adjacent courses

¼-³⁄₈ in.

Starter course

FIGURE 7-14 Wood shingles

You Should Know

The difference between wood shingles and shakes

Wood shingles

- Sawn on both sides
- Tapered
- Approximately ⅜-inch thick at butt

Wood shakes

- Split on one or both sides, or may be sawn on both sides
- Approximately ½ inch or ¾ inch thick at butt
- Usually tapered
- Greater texture and shadow lines than wood shingles ●

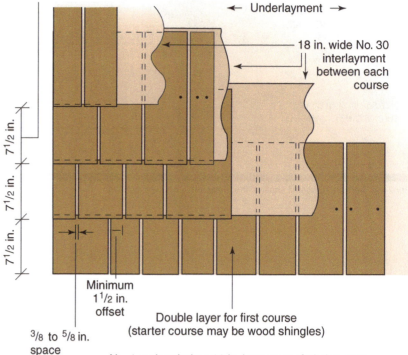

7½ in. exposure shown for 18 in. No. 1 cedar shakes on roof pitch ≥ 4:12

← Underlayment →

18 in. wide No. 30 interlayment between each course

7½ in.

7½ in.

7½ in.

Minimum 1½ in. offset

Double layer for first course (starter course may be wood shingles)

⅜ to ⅝ in. space

No. 1 cedar shakes 18 in. long on roof pitch ≥4:12

FIGURE 7-15 Wood shakes

Health and Safety

8

Home Safety

In protecting the health, safety, and welfare of the dwelling occupants, the *International Residential Code* (IRC) sets minimum requirements for a safe means of exiting the building, protection from falls and from the hazards associated with breaking glass. The code also sets minimum room dimensions to support a healthy living environment (Figure 8-1).

FIGURE 8-1 A spiral stairway is one type of stair permitted as a means of egress.

ROOM AREAS

Though most homes will far exceed the minimum requirements, the IRC recognizes the need for basic living spaces. The code requires habitable rooms other than kitchens to be 70 square feet or larger, with the smallest dimension no less than 7 feet. **[Ref. R304]**

CEILING HEIGHT

Adequate ceiling height contributes to a healthy living environment and provides the ability to move about and safely exit the building. The general rule establishes a minimum ceiling height of 7 feet for habitable space and hallways in a dwelling (Figure 8-2). The code allows for sloped ceilings, provided that half of the required room area accommodates the 7-foot height. Reductions are also permitted in basements, laundry rooms and bathrooms. **[Ref. R305]**

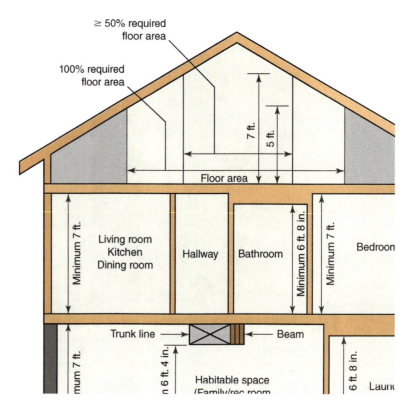

FIGURE 8-2 Ceiling height

MEANS OF EGRESS

Means of egress describes the path of travel from any location in the dwelling to the exterior. The IRC regulates stairways, ramps, hallways and doors as the primary components of that path for a safe exit from the building. Hallways must have a clear width of 3 feet, and one exterior exit door is required, also a nominal 3 feet in width. Otherwise, the IRC does not regulate the size or type of doors or limit the travel distance from any portion of the dwelling unit to the required exit. As a measure for protecting the path for safe exit from the building, the code requires limited fire resistance on the underside of stairs when the space below is enclosed. Protection is achieved by applying ½-inch gypsum board on the enclosed side. Another important principle for safe egress from the building to the outdoors is to ensure that the occupant can open the required exit door without a key or special knowledge. This precludes the use of a double-keyed deadbolt. The code does not restrict the type of hardware on other doors of the dwelling unit. **[Ref. R302.7, R311]**

Doors and landings

For each dwelling unit, the code requires one side-hinged exterior exit door providing a net opening of 32 inches by 78 inches, typically achieved with the installation of a door with nominal measurements of 3 feet by 6 feet 8 inches. Occupants in any location of the dwelling must be provided a route to the required exit door without passing through a garage. A landing or floor is generally required on each side of exterior doors with a maximum threshold height above the landing of 1½ inches. An exception allows the exterior landing at the required exit door to be not more than 7¾ inches below the top of the threshold, provided the door swings in. At other than the required exit door, the floor or landing on either side of the door is permitted to be 7¾ inches below the top of the threshold, and the door may swing in either direction (Figures 8-3 and 8-4). The code requires landings to be at least as wide as the door and not less than 36 inches in the direction of travel. A stair without a landing is permitted outside a door other than the required exit door if the door swings in and the stair has only two risers (Figure 8-5). [Ref. R311.3]

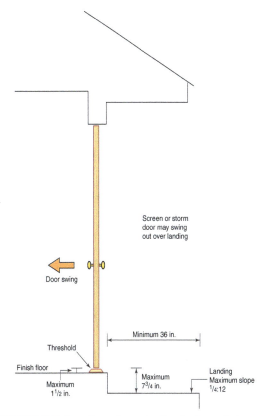

FIGURE 8-3 Landing at required exterior exit door

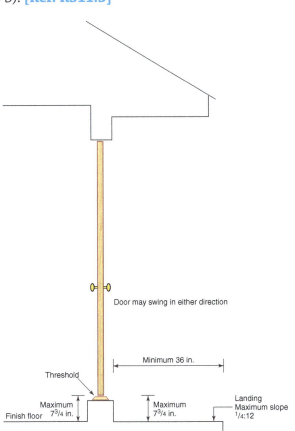

FIGURE 8-4 Landing and floor at exterior door that is not the required exit

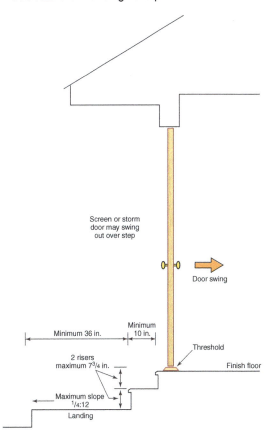

FIGURE 8-5 Steps at exterior door that is not the required exit

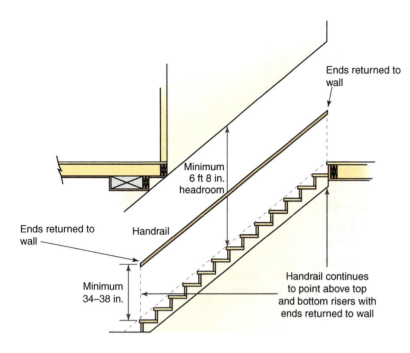

FIGURE 8-6 Stairway headroom and handrail height

Stairs

The code endeavors to improve stair safety and prevent injuries from falls by limiting the slope of the stair and by providing for minimum tread size, clearances, uniformity, and graspable handrails. The minimum 10-inch treads and maximum 7¾-inch risers determine the maximum steepness of the stairway, but just as important in stair safety is the uniformity of those treads and risers for the full flight of the stair. As a person walks a stair, he or she anticipates that the next step will be the same as the previous one. Variations that are not visually apparent may break the user's rhythm or otherwise cause a misstep and fall (Figures 8-6 through 8-8). **[Ref. R311.7]**

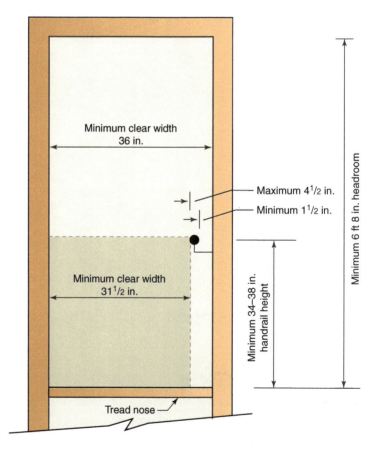

FIGURE 8-7 Stairway section, minimum width and height

Code Essentials

Prescriptive vs. performance

Prescriptive code provisions

- Handrail height is 34–38 inches
- Min. guard height is 36 inches

Performance code provisions

- Handrails shall provide equivalent graspability
- Handrails and guard top rails shall resist a single concentrated load of 200 pounds ●

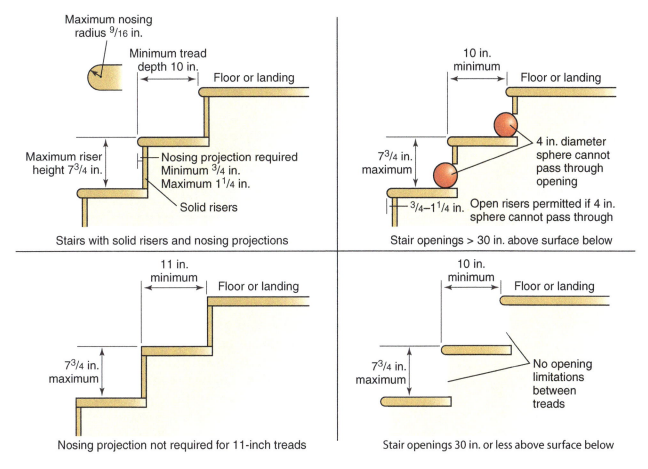

FIGURE 8-8 Stair profiles

Winders

Winder treads have nonparallel edges, and the code permits a tread depth of 6 inches at the narrow end, provided the full tread depth of 10 inches is achieved within 12 inches of the narrow side (Figure 8-9). A person walking on winder treads and holding the handrail will typically be positioned approximately 12 inches from the narrow side (referred to as the "walk line"), and this configuration allows a turn in the stairway without a landing and without creating an undue hazard. [Ref. R311.7.5.2.1]

Spiral stairways

The code permits spiral stairways at any location in a dwelling (Figure 8-10). They have lesser dimensions for width, treads and headroom and greater riser heights than conventional stairs. A spiral stairway is permitted to serve as a means of egress from one level or story to another. [Ref. R311.7.10.1]

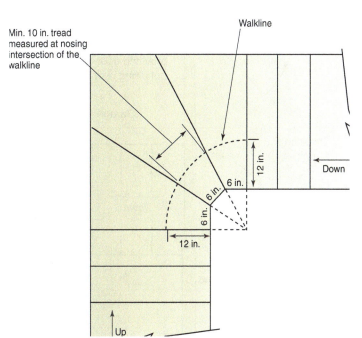

FIGURE 8-9 Winders are permitted in the same flight of stairs as rectangular treads.

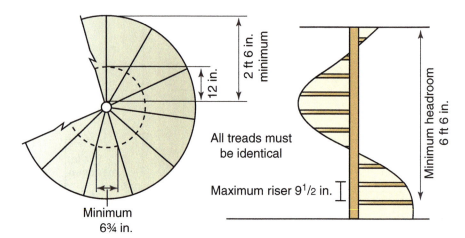

FIGURE 8-10 Spiral stair

Stair landings

Similar to the general rule requiring landings at exterior doors, a floor or landing is required at the top and bottom of stairs. This is usually not an issue in new construction where adequate space is provided between the stair and a wall or door. The landing requirement prevents the installation of a door in close proximity to the bottom tread. Such an installation would create not only a headroom problem but a falling hazard as well. An exception to the landing requirement allows a door at the top of an interior flight of stairs, provided the door does not swing over the step (Figure 8-11). **[Ref. R311.7.6]**

Handrails

Handrails are a critical component of stair safety. To be effective, they must be placed 34 to 38 inches above the tread nosing, be continuous, and have a shape that is easily grasped and held (Figures 8-12 and 8-13). For circular handrails, the IRC prescribes a diameter of 1¼ to 2 inches. The code provides dimensions for other than round handrails, but any shape, type or size of handrail that provides equivalent graspability is acceptable. This is a performance criterion, and the building official is responsible for determining equivalency. Handrails must also be securely anchored to resist a single concentrated load of 200 pounds applied in any direction. **[Ref. R311.7.8]**

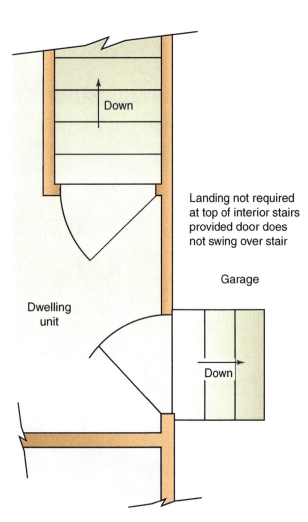

FIGURE 8-11 Door permitted at top of interior stairways

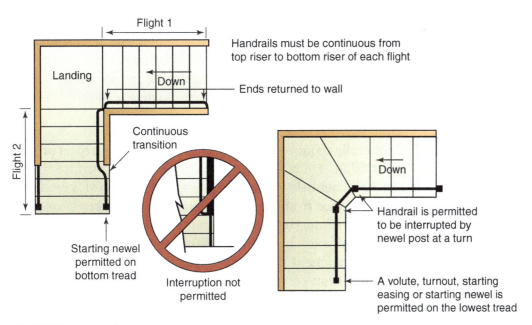

FIGURE 8-12 Handrail continuity

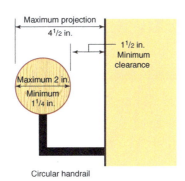

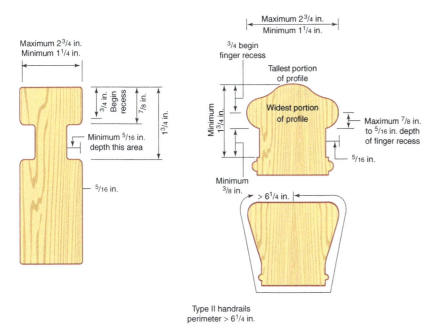

FIGURE 8-13 Handrail shapes for graspability

PROTECTION FROM FALLS

The IRC intends to protect occupants from fall injuries at prescribed locations considered hazardous by regulating the location, design and installation of guards and the height of window sills.

Guards

The IRC generally requires a minimum 36-inch-high guard as protection against falling from a walking surface to a lower surface that is more than 30 inches below. In determining where a guard is required, the vertical distance from the walking surface to the grade or floor below is measured to the lowest point within 36 inches horizontally from the edge of the open sided walking surface. The 36-inch horizontal measurement accounts for the increased hazard of a steep slope or sudden drop-off near a deck or porch. The minimum guard height is measured from the walking surface (Figure 8-14). At the sides of stairs, the minimum guard height is reduced to 34 inches to correlate with the minimum handrail height. The top rail of a stair guard often also serves as the stair handrail.

You Should Know

Wood deck guards

DCA 6—*Prescriptive Residential Wood Deck Construction Guide*

- Published by American Wood Council (AWC)
- Available at www.awc.org/codes-standards/publications/dca6
- Provides construction and connection details for wood deck guards
- Complies with the IRC
- Only one option
- Many other construction methods can be used. ●

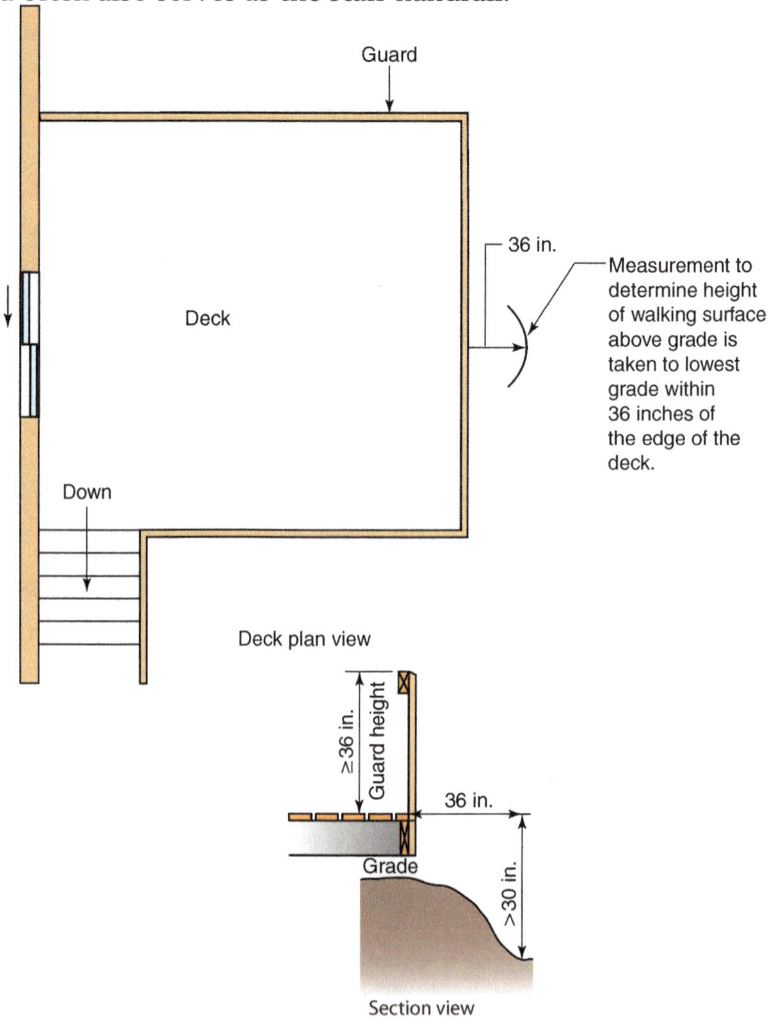

FIGURE 8-14 Determining when a guard is required at a deck

Guards also must be constructed in such a way that a 4-inch sphere will not pass through, a dimension determined after lengthy research to prevent small children from maneuvering through or becoming entrapped in such a barrier. The code grants two exceptions at the sides of stairs. The first increases the dimension to a 6-inch sphere at the triangle formed by the tread, riser, and bottom rail because of the impracticality of reducing the triangle and the negligible hazard. The second stipulates that a 4⅜-inch sphere cannot pass through a guard on the sides of stairs, a measurement that accommodates a practical wood spindle layout for staircases when building to the code-prescribed stair dimensions. Both exceptions are reasonable compromises, particularly when considered against the more common incidence of a child falling down the stair itself (Figure 8-15).

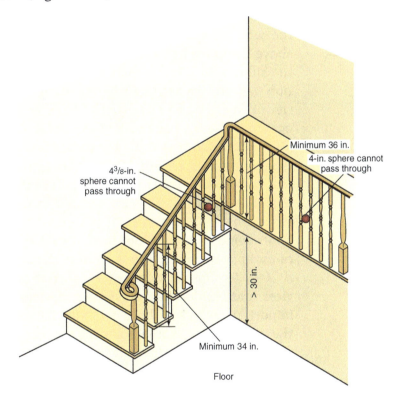

4⅜-in. sphere cannot pass through

Minimum 36 in.

4-in. sphere cannot pass through

> 30 in.

Minimum 34 in.

Floor

FIGURE 8-15 Guard at interior stair and landing open on one side

The height and allowable openings in guards are prescriptive requirements that are objectively measurable for compliance with the code. On the other hand, construction of a top rail to resist a single concentrated load of 200 pounds applied in any direction and the infill components to resist a 50-pound horizontal load applied to an area of 1 square foot are performance requirements. The adequacy of a guard or handrail to resist these loading requirements is not so easily measured or verified, though an experienced builder or inspector can fairly accurately gage the strength and stiffness of a guard or handrail assembly. For additional information on construction of guards, see the discussion of live loads in Chapter 4.

Although the most common type of guard may be constructed of wood or steel posts or balusters and a top rail, the code by no means limits the materials or methods of construction when the guard performs to the stated criteria. **[Ref. R312.1]**

Window-sill height

The minimum window-sill height requirements are intended to reduce the number of injuries to children from falls through open windows. The 24-inch sill height is typically above a small child's center of gravity, reducing the likelihood of the child's toppling over the sill. The code regulates this minimum sill height only when the window opening is more than 72 inches above the grade below. In such locations where the sill height is lower than 24 inches, protection can still be achieved by installing a barrier or limiting the dimensions of the window opening. The first alternative is a window fall prevention device meeting the requirements of the referenced standard. Likewise, an approved window opening control device conforming to the same standard may also be used to satisfy the sill height provisions. This type of device limits the opening size so that a 4-inch sphere cannot pass through. The IRC references ASTM F 2090 for both window opening control devices and window fall prevention devices. An important aspect of both types of devices is the requirement in the standard for a release mechanism for emergency escape, a provision that is in effect for either device installed at any window, whether or not the window is required to be an emergency escape and rescue opening (Figures 8-16 and 8-17). **[Ref. R312.2]**

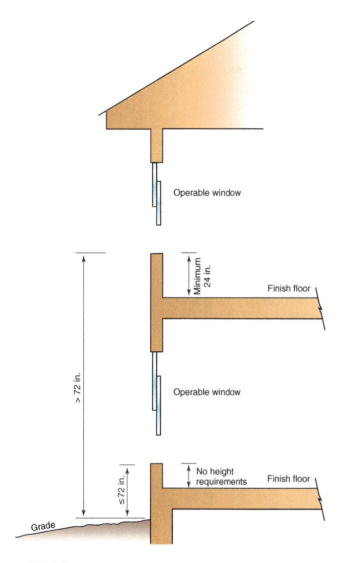

FIGURE 8-16 Window-sill height

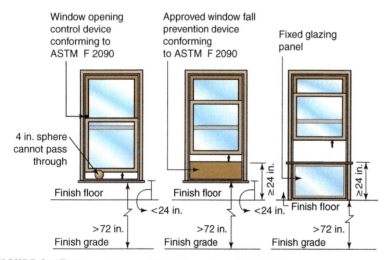

FIGURE 8-17 Alternatives to 24-inch window-sill height

EMERGENCY ESCAPE AND RESCUE OPENINGS

One of the most important safety provisions in the IRC concerns openings for emergency escape and rescue. These openings provide alternate means to escape from a sleeping room or basement in the event that a fire or other emergency blocks the usual path of egress. They allow occupants to escape directly to the safety of the outdoors and allow rescue personnel fully equipped with breathing apparatus to enter the room from the outside. Occupants are most vulnerable to the hazards of fire when they are not fully alert or when they are occupying a basement, a space that traditionally has few windows or doors and often serves as a play or recreation area. The code addresses these life-safety issues by requiring an emergency escape and rescue opening in the basement and in every sleeping room. In addition, habitable attics, which are unusual in new homes, require an emergency escape and rescue opening. For dwelling units equipped with an automatic fire sprinkler system, basement bedrooms do not require emergency escape and rescue openings when the basement provides two ways out. The first is satisfied with the usual means of egress path, typically an interior stairway. For the second, the code requires an additional means of egress or an emergency escape and rescue opening that opens directly to the outdoors (Figures 8-18 and 8-19).

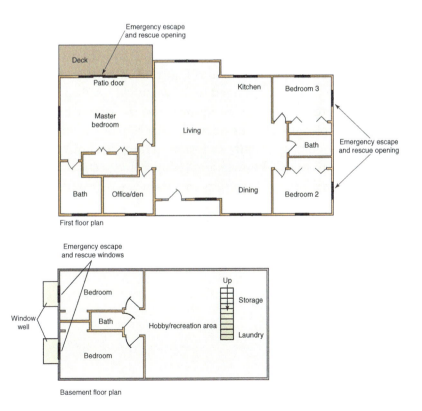

FIGURE 8-18 Required locations for emergency escape and rescue openings

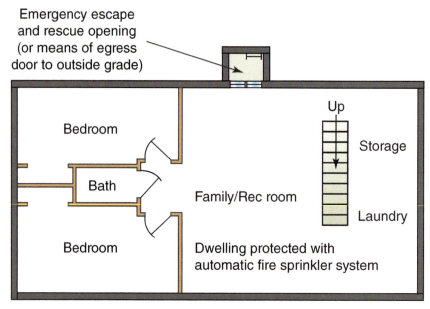

Emergency escape
and rescue opening
(or means of egress
door to outside grade)

Up

Bedroom

Storage

Bath

Family/Rec room

Laundry

Bedroom

Dwelling protected with
automatic fire sprinkler system

Basement floor plan

FIGURE 8-19 A dwelling unit protected with an automatic fire sprinkler system does not require emergency escape and rescue openings in basement bedrooms when providing two ways out of the basement.

In order for emergency escape and rescue openings to effectively serve their intended purpose, the code prescribes a maximum sill height above the floor of 44 inches and a minimum net opening size of 5.7 square feet (5.0 square feet for grade floor or below-grade openings). Width and height may be any number of combinations to achieve the minimum required opening area, provided the net width of the opening is not less than 20 inches and the net height not less than 24 inches (Table 8-1 and Figure 8-20). **[Ref. R310]**

In an emergency, occupants need to move quickly and easily to an outside space. Therefore, the code requires that the prescribed opening dimensions be obtained by the normal operation of the emergency escape and rescue opening, usually a window or door, without the need for a key, tool, or any special knowledge. This precludes the removal of a window sash or mechanical fasteners to obtain the required opening dimensions. **[Ref. R310.1.1]**

TABLE 8-1 Emergency escape and rescue windows*

Inches					
Width	20	20.5	21	21.5	22	22.5
Height	41	40	39.1	38.2	37.3	36.5
Width	23	23.5	24	24.5	25	25.5
Height	35.7	34.9	34.2	33.5	32.8	32.2
Width	26	26.5	27	27.5	28	28.5
Height	31.6	31	30.4	29.8	29.3	28.8
Width	29	29.5	30	30.5	31	31.5
Height	28.3	27.8	27.4	26.9	26.5	26.1
Width	32	32.5	33	33.5	34	34.2
Height	25.7	25.3	24.9	24.5	24.1	24

*Minimum net clear width/height combinations to obtain a net opening of 5.7 square feet.

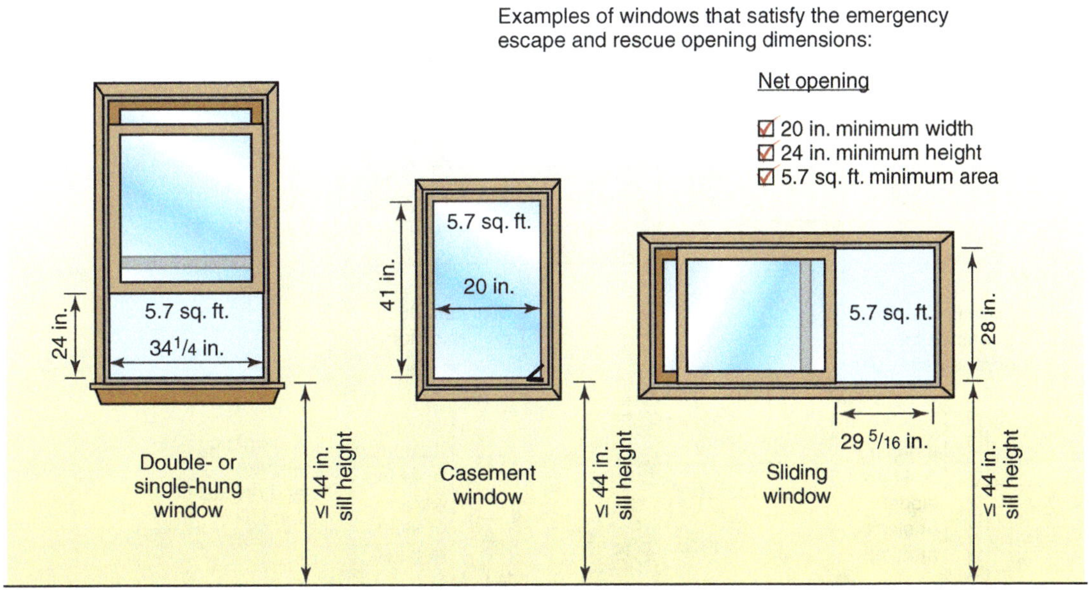

Examples of windows that satisfy the emergency escape and rescue opening dimensions:

Net opening

☑ 20 in. minimum width
☑ 24 in. minimum height
☑ 5.7 sq. ft. minimum area

Double- or single-hung window

Casement window

Sliding window

Finish floor

FIGURE 8-20 Emergency escape and rescue windows

Window wells

The IRC requires a window well when the sill of the emergency escape and rescue window is below the adjacent ground elevation. The window well area must be at least 9 square feet with a minimum dimension of 36 inches. The code requires a ladder or steps when the window well is greater than 44 inches deep, a dimension consistent with the maximum window-sill height. A ladder or step is allowed to encroach into the required window well area no more than 6 inches. Except in areas with well-drained soils, window wells serving emergency escape and rescue openings must have a means to drain to the foundation drainage system (Figures 8-21 and 8-22). [Ref. R310.2.3]

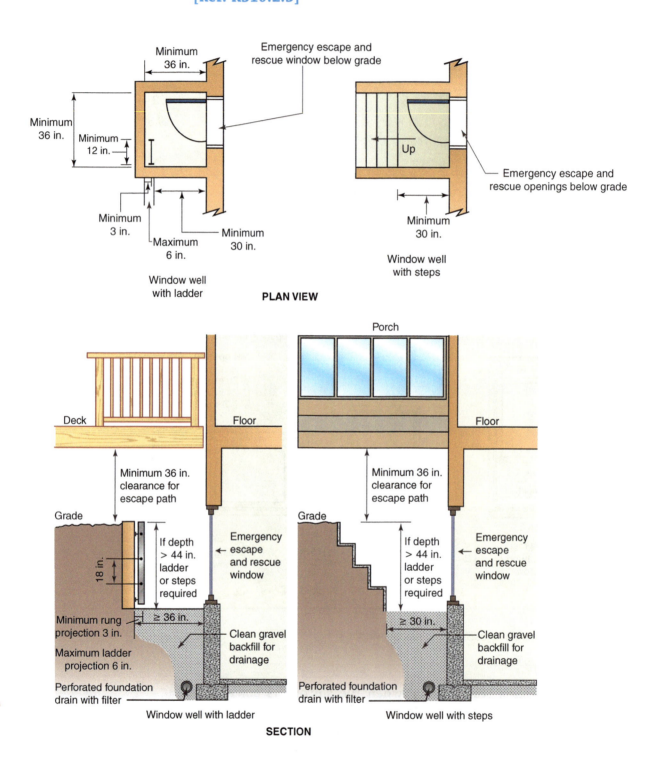

PLAN VIEW

SECTION

FIGURE 8-21 Ladder or steps required for window wells deeper than 44 inches and serving emergency escape and rescue openings

Area Wells

Similar to window wells, area wells serve below-grade doors used as emergency escape and rescue openings. The same minimum dimensions apply to doors as to windows used for this purpose. Likewise, the area well minimum width and the provisions for ladders and steps align with the corresponding window well requirements. Ladders or steps are required when the area well (or window well) is deeper than 44 inches. The code specifically exempts these ladders and steps from the means of egress provisions (Figure 8-23). [Ref. R310.3.2]

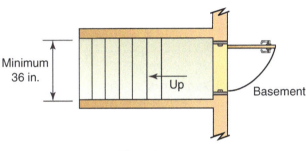

Plan view

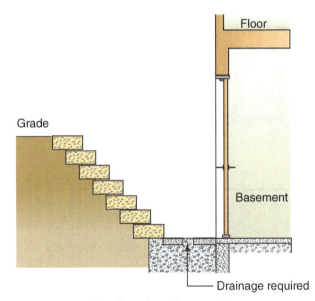

Section view of area well

FIGURE 8-22 Door serving as a required emergency escape and rescue opening from a basement with an area well

SAFETY GLASS

To prevent serious injury from shards of breaking glass, the IRC identifies seven specific locations as subject to impact by people and therefore hazardous for the installation of glazing. For example, glass in doors and adjacent to doors has an increased likelihood of accidental breakage due to actions to open and close the door and the movement of the door itself. Large panels of glass lack the visual cues or physical barriers to prevent people from accidentally walking into them. Glass adjacent to stairs is considered in a hazardous location because of the increased chances of a misstep or fall. Examples of hazardous locations subject to human impact and requiring safety glazing when glazing is installed are illustrated in Figures 8-23 through 8-29. [Ref. R308.4]

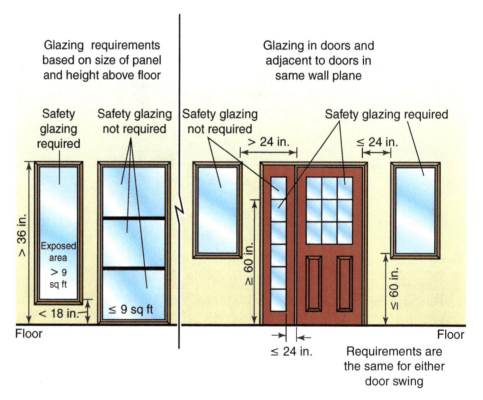

FIGURE 8-23 Safety glazing locations

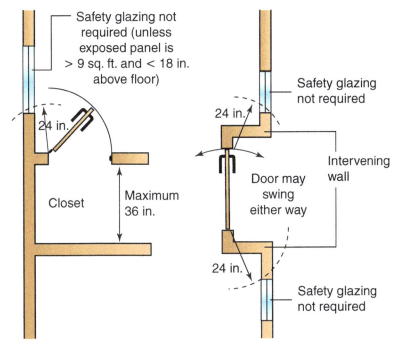

FIGURE 8-24 Glazing near doors

Yes indicates safety glazing is required

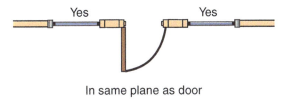

In same plane as door

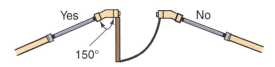

Angle less than 180 degrees from plane of door

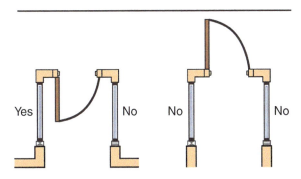

90-degree angle to plane of door

FIGURE 8-25 Glazing adjacent to doors

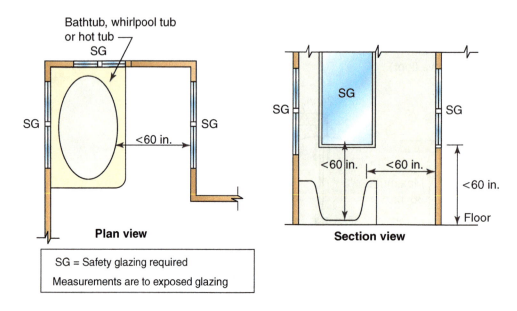

FIGURE 8-26 Glazing facing or enclosing bathtubs

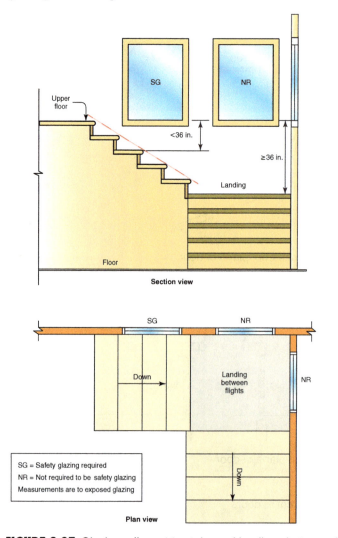

FIGURE 8-27 Glazing adjacent to stairs and landings between flights of stairs

Safety glazing, typically tempered or laminated glass, must pass the test requirements and be classified in accordance with the applicable referenced standard based on the location of the glazing. Polished wired glass is not permitted in hazardous locations requiring safety glazing, including those locations requiring fire resistance, unless it has received an approved classification through testing. The code generally requires a permanent manufacturer's designation on each panel of safety glazing indicating the type of glazing and the applicable standard. The designation is typically etched or embossed on the glass, but a label may be used, provided that it cannot be removed without being destroyed, thus preventing its reuse on a panel that is not safety glazing. For other than tempered glass, the building official may approve a certificate or affidavit of code compliance in lieu of a manufacturer's designation. **[Ref. R308.1 and R308.3]**

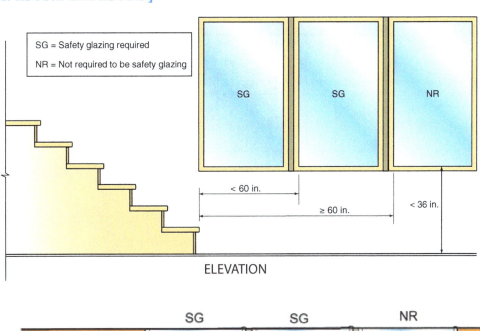

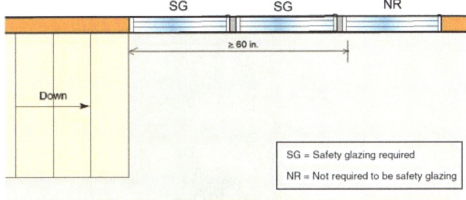

FIGURE 8-28 Glazing adjacent to the bottom landing of a stairway

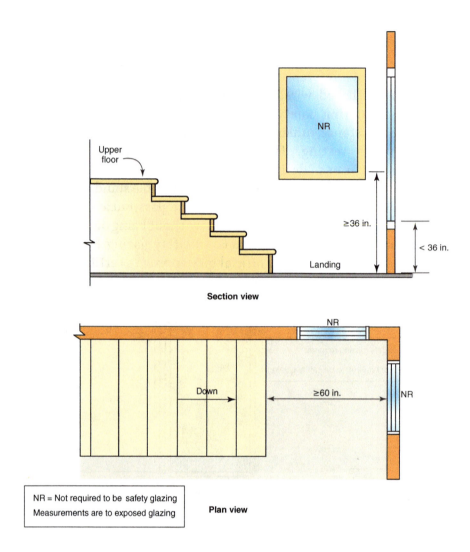

Upper floor

≥36 in.

< 36 in.

Landing

NR

Section view

NR

Down

≥60 in.

NR

NR = Not required to be safety glazing

Measurements are to exposed glazing

Plan view

FIGURE 8-29 Safety glazing not required at bottom landing when glass is at least 36 inches above floor or at least 60 inches horizontally from bottom tread

CHAPTER 9

Fire Safety

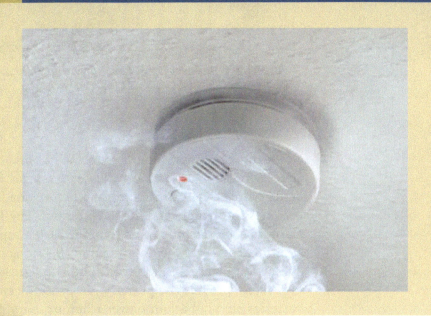

Construction under the *International Residential Code* (IRC) protects occupants from the hazards of fire through interconnected smoke alarms, an automatic fire sprinkler system, and installation of fire resistant materials at prescribed locations (Figure 9-1). Such fire protection systems are concerned primarily with life safety and provide occupants time to safely exit the building. Although smoke alarms, residential sprinkler systems and fire-resistant construction often buy time for effective fire fighting, resulting in reduced property damage, such protection is secondary to occupant safety.

FIGURE 9-1 Residential sprinkler in multipurpose loop system using PEX tubing (*Courtesy of Uponer Inc.*)

SMOKE ALARMS

Occupants are most vulnerable to the hazards of fire while sleeping. Detection and notification in the early stages of a fire provide residents with needed time to escape before the interior environment becomes intolerable and life-threatening. The IRC requires a smoke alarm in each sleeping room, outside each sleeping area and on each additional story of the dwelling unit including basements and habitable attics (Figure 9-2). The code also stipulates that the building wiring system provide the primary power to the smoke alarms and that batteries supply backup power when primary power is interrupted. [Ref. R314]

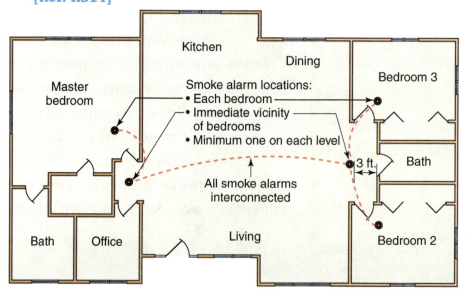

FIGURE 9-2 Smoke alarm locations

A smoke alarm is a self-contained device that provides both smoke detection and an alarm-sounding appliance. Smoke alarms must be listed as conforming to UL 217, Single and Multiple Station Smoke Alarms. The IRC requires interconnection of the devices such that when smoke is detected in one location, the alarms are activated in all locations. In addition to alerting residents on any story of the dwelling unit, interconnection ensures that the alarm is delivered to each bedroom at a sound-pressure level considered sufficient to wake a sleeping person. Wireless smoke alarms satisfy the interconnection requirements. [Ref. R314.1, R314.4]

The IRC also regulates smoke alarms for existing dwellings when interior alterations or repairs requiring a permit occur or when an addition other than a deck or porch is constructed. In this case, the residence must be brought into conformance with the location requirements for smoke alarms in new construction. However, the electrical power requirements differ from those for new buildings. Because of the costs of installing electrical wiring in existing dwellings, the code allows battery-operated alarms without a connection to the house wiring. Interconnection of smoke alarms is required in existing buildings in all cases. Because of the availability of wireless alarms, interconnection can be achieved without removing finishes or installing additional electrical wiring. The retroactive provisions for smoke alarms do not apply to exterior renovations or to the addition of a deck or porch. Plumbing or mechanical work in existing dwellings also does not trigger the smoke alarm requirements. [Ref. R314.2.2, R314.4, R314.6]

As an alternative to smoke alarms, the IRC permits a fire alarm system installed in accordance with the household fire warning equipment provisions of NFPA 72, *National Fire Alarm Code*. A household fire alarm system typically has separate devices for smoke detection and alarm annunciation. Any such system must provide detection and notification equivalent to the IRC-prescribed requirements for smoke alarms. For example, a household fire warning system may have fewer notification devices placed in the building. When a detector in any of the prescribed locations activates the system, the alarm must be clearly audible in all bedrooms over background noise levels with all intervening doors closed. In general, the sound-pressure level at the pillow cannot be less than 70 decibels. [Ref. R314.7]

FIRE SPRINKLER SYSTEMS

An automatic fire sprinkler system conforming to IRC Section P2904, Dwelling Unit Fire Sprinkler Systems, or NFPA 13D, Installation of Sprinkler Systems in One- and Two-Family Dwellings and Manufactured Homes, aids in the detection and control of fires in dwellings and intends to prevent total fire involvement (flashover) in the room of fire origin for a period of time to allow the escape or evacuation of the dwelling occupants (Figure 9-3). The IRC requires an automatic fire sprinkler system in all new one- and two-family dwellings and townhouses.

FIGURE 9-3 Residential sprinkler rough-in using approved CPVC pipe

IRC Section P2904 provides a simple, prescriptive approach to the design of dwelling fire sprinkler systems and is an approved alternative to NFPA 13D, which allows for engineered design options and other piping configurations. Consistent with structural and other design provisions in the code, the IRC prescriptive methods allow contactors, plumbers, and homeowners to design and install a dwelling sprinkler system while still providing the flexibility of an engineered design in accordance with NFPA 13D.

Although the title of NFPA 13D suggests that the application of the standard is limited to one- and two-family dwellings, it is also appropriate for automatic sprinkler systems installed in townhouses regulated by the IRC. Each townhouse has a separate connection to water, wastewater, fuel-gas, and electrical utilities, and such connections are under the control of the occupant. Conversely, an NFPA 13R system is designed for the protection of apartment buildings, lodging and rooming houses, and hotels, motels, and dormitories up to four stories in height. In these occupancies, the property management is responsible for the maintenance of the entire building in accordance with the IBC, the International Fire Code (IFC), and the International Property Maintenance Code (IPMC).

A dwelling fire sprinkler system requires less water when compared to NFPA 13 and 13R systems. Section P2904 and NFPA 13D generally require a minimum water discharge duration of 10 minutes, compared to 30 minutes for an NFPA 13R system and even higher duration values in NFPA 13. The minimum sprinkler discharge density may be satisfied by connection to a domestic water supply, a water well, an elevated storage tank, an approved pressure tank or a stored water source with an automatically operated pump. Any combination of water supply systems is allowed to meet the required dwelling fire sprinkler system capacity.

Unlike an NFPA 13 automatic sprinkler system, a dwelling fire sprinkler system does not require automatic sprinkler protection throughout a one- and two-family dwelling or townhouse. Sprinklers are not required in areas that have been statistically shown through fire incident loss data to not significantly contribute to injuries or death. For example, sprinklers may be omitted in closets with areas of 24 square feet or less and bathrooms with areas of 55 square feet or less. Generally, sprinklers also are not required in open attached porches, garages, attics, crawl spaces and concealed spaces not intended or used for living purposes. In addition, a dwelling fire sprinkler system does not require a fire department connection.

The provisions for dwelling fire sprinkler systems require the use of new sprinklers listed for residential applications. These sprinklers are equipped with a thermal element using either a fusible link or frangible bulb filled with conductive liquid that is designed to operate approximately five times faster when compared to a standard spray sprinkler required by NFPA 13 installed in the same setting. The sprinkler discharge pattern is designed to wet the walls and floors of the space and apply water onto any furnishings. [Ref. R313 and P2904]

FIRE-RESISTANCE-RATED CONSTRUCTION

Fire-resistance-rated construction limits the spread of fire to protect property and occupants, and provides time for effective fire-fighting efforts. The components of the wall or floor-ceiling construction form an assembly that has proven through testing to resist the effects of fire for the designated time period. The fire-resistance rating is based on results under the specific test conditions and does not necessarily predict performance under actual field conditions.

Occupants have no control over the action of their neighbors, and the fire-resistant-rated construction of exterior walls located near lot lines and the separation between dwelling units offers an appropriate level of protection.

Exterior walls

To protect against the spread of fire from a building on one property to a building on another property, the code prescribes an intervening clear space between buildings and the lot line or requires fire-resistant construction of exterior walls. When exterior walls are located less than the prescribed distance from the lot line (5 feet for buildings without fire sprinkler protection and 3 feet for buildings with fire sprinkler protection), the IRC requires a fire-resistance rating of one hour. (Figure 9-4).

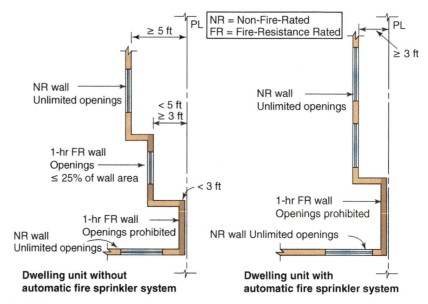

FIGURE 9-4 Fire resistance of exterior walls adjacent to lot lines

The IRC also regulates projections and openings within this setback zone of 3 to 5 feet. Openings, typically windows or doors, are not permitted less than 3 feet from the lot line. For dwellings without sprinklers and detached accessory buildings, openings that are at least 3 feet but less than 5 feet from the lot line are limited to 25 percent of the wall area. This area limitation does not apply to dwellings protected with an automatic fire sprinkler system. Foundation vents for ventilation of underfloor spaces are not regulated for minimum fire separation distance. Projections, typically roof overhangs, require one-hour fire protection on the underside when less than the prescribed distance from the lot line (5 feet without sprinklers and 3 feet with sprinklers) unless fireblocking is installed between the top of the rated wall and the roof sheathing. Roof overhangs typically cannot project closer than 2 feet from the lot line. An exception permits a garage located within 2 feet of the lot line to have a 4-inch roof eave projection (Table 9-1 and Figures 9-5 and 9-6). For further discussion of exterior walls and fire separation distances, see Chapter 3. **[Ref. R302.1]**

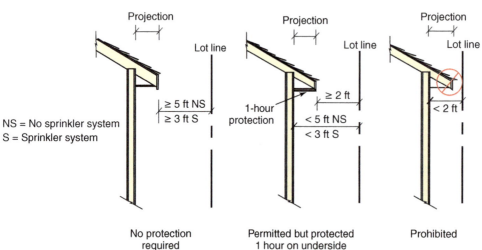

FIGURE 9-5 Roof projections at property line

TABLE 9-1 Fire resistance of exterior walls adjacent to lot lines

Exterior wall element	Fire separation distance to lot line	Dwellings without sprinkler system and detached accessory buildings[1]		Dwellings with sprinkler system	
		Minimum fire-resistance rating	Restrictions	Minimum fire-resistance rating	Restrictions
Walls	5 feet and greater	0 hours		0 hours	
	3 feet to < 5 feet	1 hour	Tested for exposure from both sides	0 hours	
	0 feet to < 3 feet	1 hour	Tested for exposure from both sides	1 hour	Tested for exposure from outside
Projections	5 feet and greater	0 hours		0 hours	
	3 feet to < 5 feet	1 hour	On the underside	0 hours	
	2 feet to < 3 feet	1 hour	On the underside	1 hour	On the underside
	0 feet to < 2 feet	Not permitted[2]		Not permitted	
Openings in walls	5 feet and greater	0 hours	Unlimited area	0 hours	Unlimited area
	3 feet to < 5 feet	0 hours	Maximum 25% of wall area	0 hours	Unlimited area
	0 feet to < 3 feet	Not permitted		Not permitted	

1. Detached sheds not greater than 200 square feet in area are not regulated for separation distance.

2. Detached garages within 2 feet of lot line are permitted to have a 4-inch roof eave overhang.

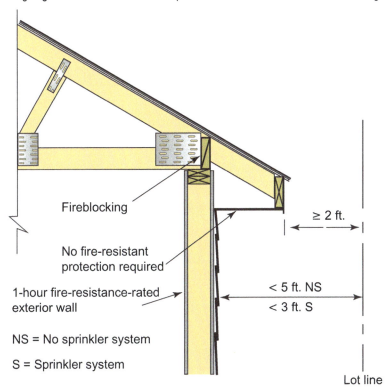

FIGURE 9-6 A fire-resistance rating is not required for roof eave projections when fireblocking is installed above the exterior wall.

Two-family dwellings

The IRC requires a one-hour fire-resistance-rated separation between the dwelling units of a two-family dwelling. Horizontal floor-ceiling assemblies separating upper and lower units must extend to the exterior walls, and supporting wall construction must also be one-hour fire-resistance-rated (Figure 9-7). Wall assemblies separating side-by-side units must generally extend from the foundation through the attic space to the bottom of the roof sheathing (Figure 9-8). As an alternative, the IRC permits the wall assembly to terminate at a ⅝-inch gypsum board ceiling when a draft stop is installed in the attic area and not less than ½-inch gypsum board is installed on the walls supporting the ceiling (Figure 9-9). [Ref. R302.3]

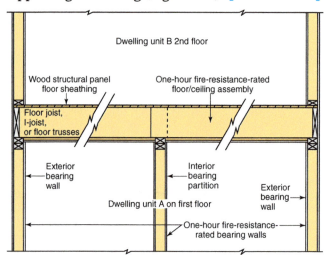

FIGURE 9-7 Horizontal separation for two-family dwelling

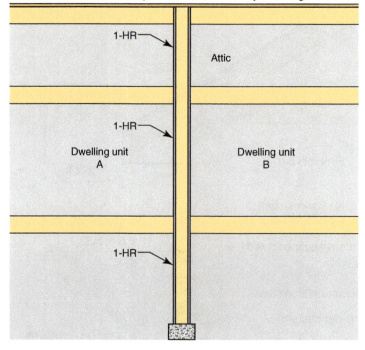

FIGURE 9-8 Two-family dwelling separation wall

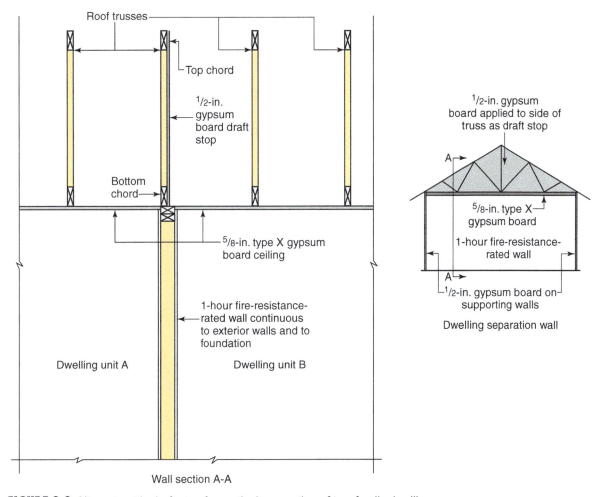

FIGURE 9-9 Alternate attic draft stop for vertical separation of two-family dwelling

Townhouses

As with two-family dwellings, a priority of the code is to separate townhouse dwelling units with fire-resistant-rated wall assemblies to limit the spread of fire in the structure and to provide some protection to occupants from events that occur in the neighboring unit. Townhouses, by definition, are single-family dwelling units constructed in a group of three or more connected units. Because the total number of townhouse dwelling units in a single building is unlimited, the IRC treats their separation somewhat differently than duplexes to provide greater protection from the spread of fire. There are two paths to achieve compliance with the code for fire-resistant separation of townhouse dwelling units. The first method is to construct two one-hour-rated wall assemblies. In this case, each townhouse is required to be structurally independent as it relates to fire resistance. The second option permits construction of a common wall between dwelling units that does not require structural independence of the townhouse units. The fire-resistance rating of the common wall between townhouses is tied to the presence of an automatic fire sprinkler system. A 2-hour rating is required for non-sprinklered

Code Essentials

For townhouses without parapets, roof penetrations are prohibited within 4 feet of the separating wall

- Skylights
- Roof windows
- Gas vents
- Plumbing vents
- Exhaust outlets
- Air intakes
- Ridge vents
- Roof vents ●

buildings. For townhouse dwelling units protected with an automatic sprinkler system, a 1-hour rating is required (Figures 9-10 and 9-11). To preserve structural integrity and limit penetrations of the fire-resistant membrane, the code does not permit the installation of plumbing or mechanical equipment, ducts or vents in the cavity of the common wall. Electrical installations are permitted and must meet the penetration requirements for fire-resistance-rated assemblies. **[Ref. R302.2]**

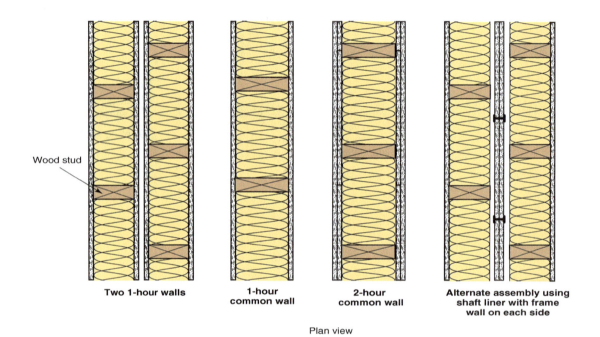

Wood stud

Two 1-hour walls   **1-hour common wall**   **2-hour common wall**   **Alternate assembly using shaft liner with frame wall on each side**

Plan view

FIGURE 9-10 Typical fire-resistant-rated wall assemblies for separating townhouse dwelling units

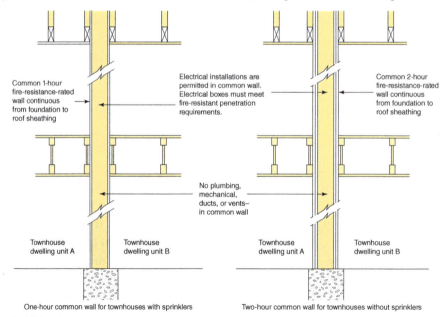

Common 1-hour fire-resistance-rated wall continuous from foundation to roof sheathing

Electrical installations are permitted in common wall. Electrical boxes must meet fire-resistant penetration requirements.

Common 2-hour fire-resistance-rated wall continuous from foundation to roof sheathing

No plumbing, mechanical, ducts, or vents— in common wall

Townhouse dwelling unit A   Townhouse dwelling unit B   Townhouse dwelling unit A   Townhouse dwelling unit B

One-hour common wall for townhouses with sprinklers   Two-hour common wall for townhouses without sprinklers

FIGURE 9-11 Common walls separating townhouses

To prevent the spread of fire from one unit to the next at the roof line, the IRC provides two options. The first requires a 30-inch-high parapet wall with equivalent fire rating as the separation wall. The second option is the more common practice of installing the prescribed fire-resistant protection at the roof for a distance of 4 feet on each side of the separating wall. This protection is not a fire-resistant-rated assembly, but the materials specified perform satisfactorily in resisting the spread of fire. Because any penetration of this protected area of the roof would reduce the effectiveness in preventing the spread of fire, the code prohibits openings or penetrations in the roof within 4 feet of the separation wall (Figures 9-12 and 9-13). [Ref. R302.2.2]

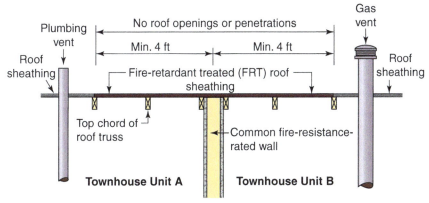

FIGURE 9-12 Townhouse roof protection on each side of separation wall

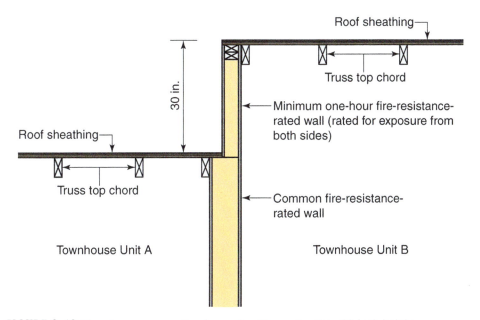

FIGURE 9-13 Townhouse separation for roofs with greater than 30-inch height difference

Fire-resistance-rated assemblies

Many tested fire-resistance-rated assemblies are available utilizing various materials and methods of construction. In buildings regulated by the IRC, gypsum board applied to wood framing is the most common type of assembly. Assemblies are assigned an hourly fire-resistance rating through testing in accordance with ASTM E 119, Test Methods for Fire Tests of Building Materials and Construction, or UL 263, Fire Tests of Building Construction and Materials. Tested assemblies are available in the *Gypsum Association Fire Resistance Design Manual* and from approved testing agencies. In addition, the code permits construction of fire-resistant-rated assemblies in accordance with Section 703.3 of the *International Building Code* (IBC). Construction must match the design specifications for types of materials, dimensions and methods of attachment (Figure 9-14). **[Ref. R302.1, R302.2, R302.3]**

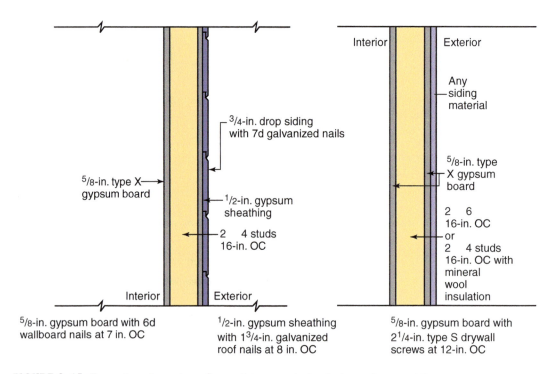

FIGURE 9-14 Examples of one-hour fire-resistance-rated exterior wall assemblies

Penetrations of fire-resistance-rated assemblies

When items such as pipes or ducts penetrate one or both sides of the fire-resistance-rated wall or floor-ceiling assembly separating dwelling units, both the penetrating item and the space around it must be protected to maintain the integrity of the fire-resistant assembly. In general, penetrations by metal pipe require that the space around the pipe be filled with approved materials to prevent the passage of flame and hot gases. The material, such as a listed fire-stop sealant, must be installed according to the manufacturer's instructions to provide a fire-resistant time rating that is equivalent to that of the construction being penetrated. Other penetrating materials, such as plastic pipe (if permissible—plumbing piping is not permitted to penetrate or be located in the common wall separating townhouses), must be protected by an approved penetration fire-stop system. Such a system often consists of intumescent material that expands when heated by fire conditions, filling the penetration as the plastic pipe melts and preserving the fire-resistance rating of the wall or floor-ceiling assembly. [Ref. R302.4]

Steel electrical boxes up to 16 square inches and listed electrical boxes are permitted to penetrate the membrane of a fire resistance rated wall assembly. Steel box penetrations are limited to 100 square inches in any 100 square feet of wall area. When electrical boxes are installed on opposite sides of the wall assembly, they require a minimum horizontal separation distance or another approved means of separation. The installation of wood blocking between the boxes or application of a listed putty pad to each box satisfies the separation requirement. Putty pads consist of intumescent fire-stop material that expands when heated to seal off the electrical box penetration (Figure 9-15). [Ref. R302.4.2]

Note: Steel boxes limited to 16 sq. in. each and total of 100 sq. in. in any 100 sq. ft. of wall

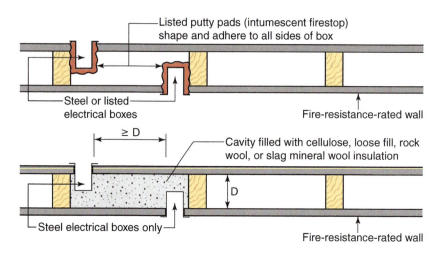

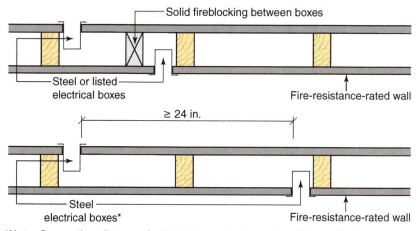

*Note: Separation distance for listed boxes is determined by the listing.

FIGURE 9-15 Electrical box penetrations on opposite sides of fire-resistance-rated wall assembly

DWELLING SEPARATION FROM GARAGE

Unlike separations between dwelling units, the separation between the residence and garage is not a fire-resistance-rated assembly. Likewise, penetrations through the separation are not required to meet the rated penetration requirements for fire-resistance-rated assemblies. Attached garages and detached garages within 3 feet of the dwelling require installation of gypsum board on the garage side to provide limited resistance to the spread of fire. Generally, the IRC prescribes ½-inch gypsum board installed on the garage side to achieve this separation (Figures 9-16 and 9-17). When there are habitable rooms above the garage, the code requires the installation of ⅝-inch Type X gypsum board on the garage ceiling. The bearing walls supporting the ceiling framing in this instance also require the application of ½-inch gypsum board on the interior surface (Figure 9-18). **[Ref. R302.6]**

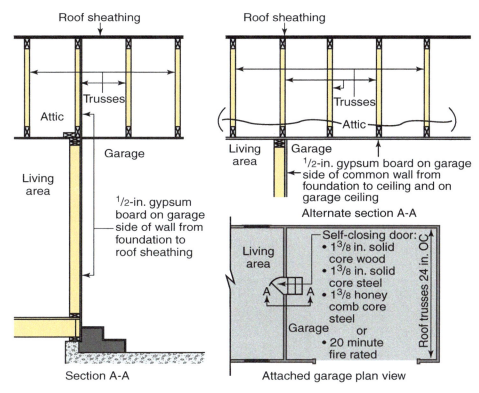

FIGURE 9-16 Dwelling and attached garage separation

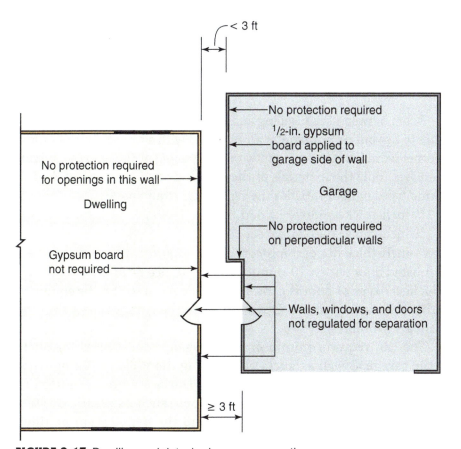

FIGURE 9-17 Dwelling and detached garage separation

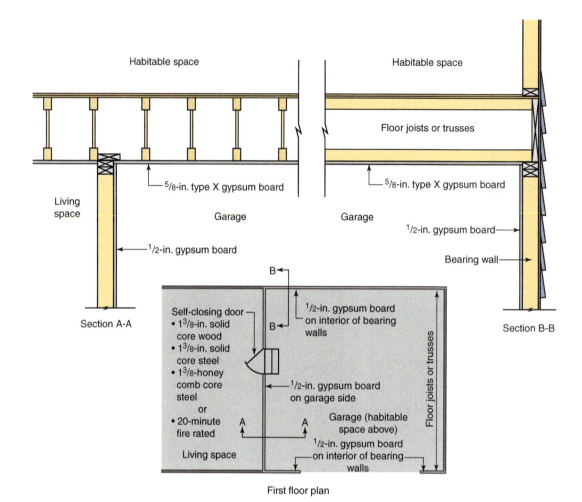

FIGURE 9-18 Separation for attached garage with habitable space above

Doors between the dwelling and the garage also provide some resistance to fire but do not require an assembly with a fire-resistance rating. In other words, the frame, hardware, and sealing of the opening are not addressed, only the materials of the door leaf itself. Any one of the following types of door satisfies the separation requirement:

- 1⅜-inch-thick solid-core wood
- 1⅜-inch-thick solid-core steel
- 1⅜-inch-thick honeycomb-core steel
- A listed door with a 20-minute fire resistance rating.

Whichever type of door is chosen, the code requires it to be equipped with a self-closing device. Openings from the garage into a sleeping room are prohibited.

The IRC requires minimum No. 26-gauge sheet steel for ducts in the garage as well as ducts penetrating the walls or ceilings that separate the dwelling from the garage. Ducts are not permitted to open into the garage. Other penetrations, such as plastic or steel pipe, require only that the space around the penetration be filled with approved materials to limit the free passage of fire and smoke. **[Ref. R302.5]**

FIRE PROTECTION OF FLOORS

Installation of ½-inch gypsum board, ⅝-inch wood structural panel or other approved material is required on the underside of certain floor assemblies of dwelling units and accessory buildings constructed under the IRC. The application of gypsum wallboard or other approved material intends to provide some protection to the floor system against the effects of fire and delay collapse of the floor, primarily as a safeguard for fire fighters. This provision applies to light-frame construction consisting of I-joists, manufactured floor trusses, cold-formed steel framing and other materials and manufactured products considered most susceptible to collapse in a fire. Solid-sawn lumber and structural composite lumber perform fairly well in retaining adequate strength under fire conditions. Therefore, floors framed with nominal 2 x 10 lumber or larger of these materials are exempt from this section's fire protection requirements. Similarly, if sprinklers are installed to protect the space below the floor assembly, additional protection is not required. Crawlspaces without storage or fuel-fired appliances are not considered to contain sufficient fuel load to present an undue hazard to floor collapse. The code also exempts small areas of ceiling, such as may occur in a utility room in a basement, from the fire protection requirements, provided the space is not open to other portions of the floor system. Therefore, fireblocking is required to isolate the unprotected area from the protected area of the floor system (Figure 9-19). [Ref. R302.13]

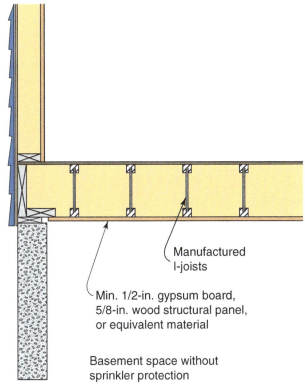

Manufactured
I-joists

Min. 1/2-in. gypsum board,
5/8-in. wood structural panel,
or equivalent material

Basement space without
sprinkler protection

FIGURE 9-19 Fire protection of floors

FOAM PLASTIC

In addition to meeting the maximum flame spread and smoke-developed indices, foam plastic insulation must be isolated from exposure to fire from the dwelling interior by the installation of a thermal barrier. The IRC prescribes the application of ½-inch gypsum board or an approved material providing equivalent protection to separate the foam plastic from the interior (Figure 9-20). In attics and crawl spaces entered only for maintenance or repairs, the IRC permits a reduction in protection, and the foam plastic may be covered with one of the prescribed ignition barrier materials (Figure 9-21). [Ref. R316.4, R316.5]

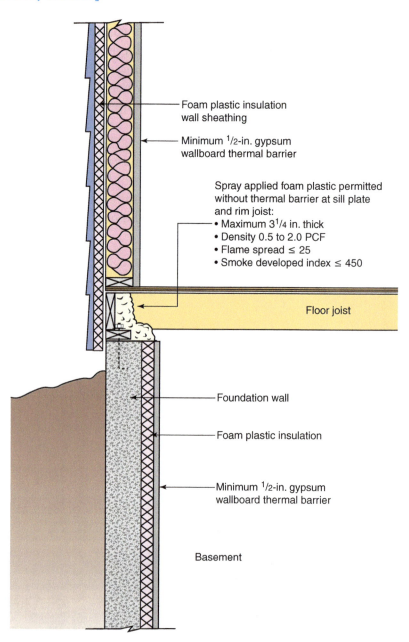

FIGURE 9-20 Foam plastic thermal barrier

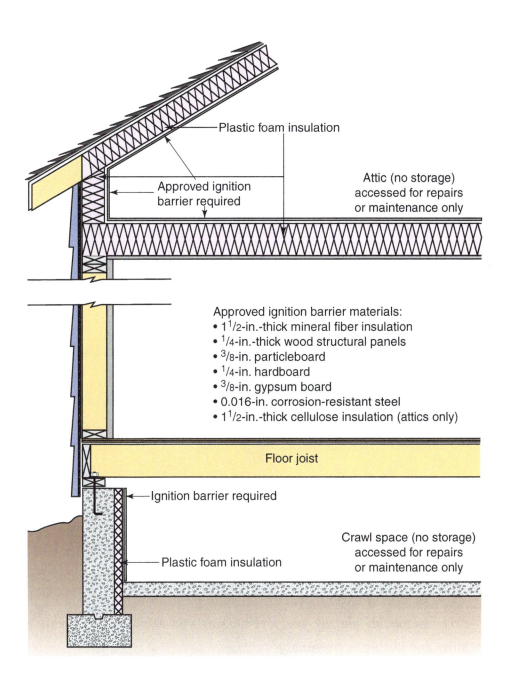

Plastic foam insulation

Approved ignition barrier required

Attic (no storage) accessed for repairs or maintenance only

Approved ignition barrier materials:
- $1\frac{1}{2}$-in.-thick mineral fiber insulation
- $\frac{1}{4}$-in.-thick wood structural panels
- $\frac{3}{8}$-in. particleboard
- $\frac{1}{4}$-in. hardboard
- $\frac{3}{8}$-in. gypsum board
- 0.016-in. corrosion-resistant steel
- $1\frac{1}{2}$-in.-thick cellulose insulation (attics only)

Floor joist

Ignition barrier required

Plastic foam insulation

Crawl space (no storage) accessed for repairs or maintenance only

FIGURE 9-21 Foam plastic ignition barrier

CHAPTER
10

Healthy Living Environment

The *International Residential Code* (IRC) sets minimum requirements for natural or artificial light, fresh air ventilation, carbon monoxide alarms, comfort heating and sanitation to create a healthy and livable environment.

NATURAL AND ARTIFICIAL LIGHT

Though the code retains the traditional standards for natural light from windows, electric lighting satisfies the minimum illumination requirements for habitable rooms in almost all cases. The minimum average illumination level for artificial lighting in habitable rooms is 6 footcandles, far below typical indoor illumination levels and lighting industry recommendations of 50 footcandles or more. Although windows may be eliminated for lighting purposes, they may still be required for emergency escape and rescue and fresh air ventilation purposes. [Ref. R303.1]

Stairway illumination

As part of the egress path and a component presenting increased hazards of fall injuries, stairway design and construction, including adequate illumina-tion, is particularly important to safety in a dwelling. The IRC requires a min-imum illumination level of 1 footcan-dle at treads and landings of interior stairs. For other than continuous or automatic il-lumination (such as provided with motion sensors), interior stairways with six or more risers require a wall switch at each floor level.

Exterior stairs require a light source located near the top landing. In addition, bottom landings require a light source if they provide access to the basement from grade level (Figure 10-1). [Ref. R303.7, R303.8, E3903.3]

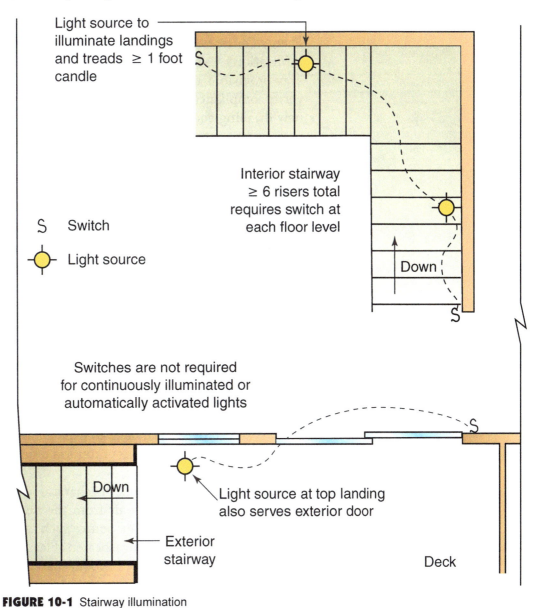

FIGURE 10-1 Stairway illumination

NATURAL AND MECHANICAL VENTILATION

To provide fresh air ventilation to habitable rooms, the IRC requires openings to the outdoors with an area equal to four percent of the floor area of the room or an approved whole-house mechanical ventilation system. Fresh air for natural ventilation is typically provided by operable windows and doors, and most dwellings have adequate openings to satisfy this requirement. However, even for dwelling units with adequate natural ventilation openings, the code requires a whole-house ventilation system when testing with a blower door under the prescribed criteria shows an air infiltration rate of less than five air changes per hour (ACH) at the specified test pressure. Because a reduced air infiltration rate is desirable and effective for conserving energy, new construction complying with the energy provisions of the IRC, which are extracted from the *International Energy Conservation Code* (IECC), is sufficiently tight to have an air infiltration rate below the threshold of five ACH. In addition, the energy provisions require a blower door test for new dwelling units to verify compliance with the energy conservation measures. Therefore, new dwelling construction typically requires a whole-house ventilation system capable of providing adequate outdoor air to maintain satisfactory indoor air quality under closed-house conditions. **[Ref. R303.1, R303.4]**

Whole-house mechanical ventilation simply exchanges outdoor air for indoor air at the minimum air-flow rates prescribed in the IRC mechanical provisions based on the area of the dwelling and the number of bedrooms. The code does not require a separate system, but permits a combination of supply and exhaust fans in achieving adequate ventilation. For example, outdoor air introduced into the return side of the heating, ventilating and air conditioning (HVAC) system is permitted to satisfy the supply air requirements. For additional information on mechanical ventilation systems, see Chapter 12. **[Ref. M1507]**

In addition to habitable rooms, bathrooms and toilet rooms require natural or mechanical ventilation. Unless windows provide 1.5 square feet of total openable area for outside air, one or more exhaust fans must be provided to exhaust air directly to the outside. Minimum mechanical exhaust rates are found in the IRC mechanical provisions. **[Ref. R303.3]**

Openable windows, doors, mechanical ventilation air intakes and similar openings that draw air from the outside must be located at least 10 feet from any noxious source such as plumbing vents, gas flue vents and chimneys. Openings located at least 3 feet below the noxious source are not subject to the 10-foot separation requirement. Exhaust and intake openings must be protected on the outside with corrosion-resistant screens with openings of ¼ to ½ inch (Figure 10-2). **[Ref. R303.5, R303.6]**

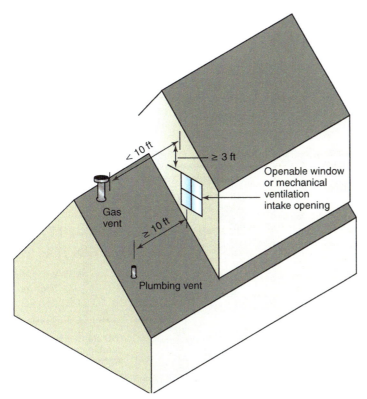

≥ 3 ft

< 10 ft

Openable window
or mechanical
ventilation
intake opening

Gas
vent

≥ 10 ft

Plumbing vent

FIGURE 10-2 Outside air intake opening locations

CARBON MONOXIDE ALARMS

As part of a safe and healthy interior living environment, the IRC provides for early warning to alert occupants to hazardous levels of carbon monoxide gas. The carbon monoxide alarm provisions are only in effect when the dwelling contains fuel-fired appliances or has an attached garage that communicates with the dwelling unit. A malfunctioning fuel-fired appliance, such as a gas-fired furnace, water heater or fireplace, is the most common cause of carbon monoxide poisoning in homes. Automobile exhaust migrating into the home from an attached garage is the other hazard addressed by the code requirements, but only if there is a door or other opening from the garage into the house. Attached garages that do not communicate with the house do not trigger the carbon monoxide alarm requirements.

The code requires carbon monoxide alarms to be installed outside of each separate sleeping area in the immediate vicinity of the bedrooms to protect people when they are most vulnerable to the effects of carbon monoxide poisoning—when they are sleeping or not fully alert. The IRC requires an additional alarm to be located in a bedroom when a fuel-fired appliance is installed in the bedroom or the adjoining bathroom (Figure 10-3). Connection to the house wiring system with battery backup is required for carbon monoxide alarms installed in new dwellings. Where two or more carbon monoxide alarms are installed to satisfy the location provisions, the code requires interconnection of the alarms so that when one alarm is activated all devices sound the alarm.

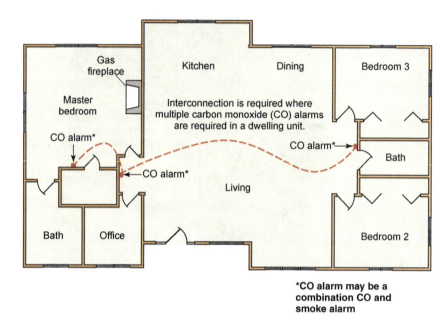

*CO alarm may be a combination CO and smoke alarm

FIGURE 10-3 Carbon monoxide (CO) alarm required in the immediate vicinity of each sleeping area and in bedrooms containing fuel-fired appliances

Similar to the smoke alarm provisions, the provisions for carbon monoxide alarms state that when interior work or an addition requiring a permit occurs, carbon monoxide alarms must be installed in the locations prescribed by the code. Exterior work, such as roofing, siding, windows, doors, and porch and deck additions, does not trigger the retroactive carbon monoxide alarm requirements. This language mirrors the exemption in the smoke alarm provisions. Battery-operated carbon monoxide alarms are permitted for satisfying the requirements in existing buildings.

Combination carbon monoxide and smoke alarms complying with the applicable standards are permitted and are commonly installed outside of bedroom areas in residential construction as an acceptable method for satisfying both smoke alarm and carbon monoxide alarm provisions in the IRC. **[Ref. R315]**

HEATING AND COOLING

In geographical areas where the design temperature is less than 60°F, as indicated by the governing authority in the adopting ordinance (see Chapter 4 of this publication), the IRC requires a heating system capable of maintaining a minimum temperature of 68°F at 3 feet above the floor and 2 feet from exterior walls. The code does not require the installation of an air conditioning or comfort cooling system. When mechanical equipment for heating or cooling is installed, it must comply with the mechanical and fuel-gas provisions of the IRC (see Chapter 12 of this publication). **[Ref. R303.9]**

SANITATION

In the building planning chapter of the code, the IRC establishes basic requirements for bathroom and kitchen fixtures, clearance dimensions, hot and cold water, and sewer connections. Installation must also comply with the specific requirements of the IRC plumbing provisions (see Chapter 13 of this publication).

Toilet and bathing facilities

In order to maintain a healthy and sanitary living environment, a residence must provide facilities for toilet, bathing and handwashing purposes. The IRC requires at least one water closet, one lavatory, and a bathtub or shower in every dwelling unit. Each fixture must be connected to an approved water supply and sewer. Lavatories, bathtubs, showers, and bidets require connection to both hot and cold water supply. [Ref. R306]

The IRC prescribes minimum clearance dimensions around bathroom fixtures so that occupants can reasonably access and use the fixtures. The minimum size of a shower is also set at 30 inches by 30 inches, though the IRC plumbing provisions provide an alternative for a narrower shower compartment with a greater area. In this case, the minimum inside width of the shower compartment is 25 inches, and the minimum inside area is 1,300 square inches, which correlates to the approximate inside dimensions of a standard bathtub. These alternative dimensions are useful in remodeling situations where a shower replaces a bathtub (Figures 10-4 and 10-5). [Ref. R307.1, P2705.1, P2708.1]

For ease of cleaning and to protect the structure from deterioration, showers must have nonabsorbent wall surfaces to a minimum height of 6 feet above the floor. [Ref. 307.2]

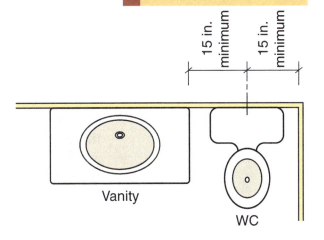

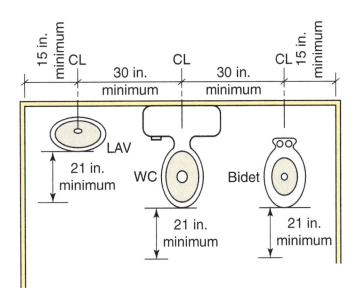

Installation of fixtures

FIGURE 10-4 Bathroom fixture clearances

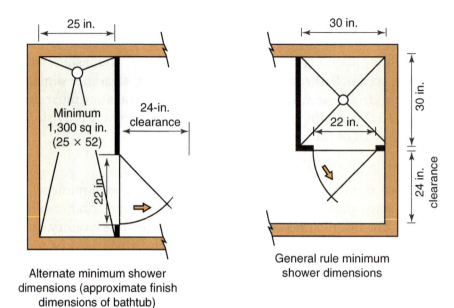

Alternate minimum shower
dimensions (approximate finish
dimensions of bathtub)

General rule minimum
shower dimensions

FIGURE 10-5 Minimum shower dimensions

Cooking and cleaning facilities

Also important to the health of the occupants are adequate and
sanitary means to prepare meals, wash dishes, and wash clothes. The
IRC requires a kitchen area and kitchen sink for every dwelling unit.
Kitchen sinks, laundry tubs and washing machine outlets require
connection to hot and cold water supply. **[Ref. R306.2, R306.4]**

Chimneys and Fireplaces

The *International Residential Code* (IRC) contains prescriptive provisions for the construction of masonry chimneys and fireplaces, including requirements for combustion air supply, clearance to combustibles and hearth construction (Figure 11-1). The appliance listing and manufacturer's instructions typically govern the installation of approved factory-built fireplaces and chimneys.

FIGURE 11-1 A masonry chimney is constructed of solid masonry units, hollow masonry units grouted solid, stone or concrete.

EXTERIOR AIR SUPPLY

Both factory-built and masonry fireplaces require an adequate exterior air supply to assure proper fuel combustion and prevent depleting oxygen within the habitable space. Mechanical ventilation of the room is permitted as an alternative when controlled so that the indoor pressure of the room or space is neutral or positive. Exterior combustion air ducts or passageways for masonry fireplaces must be at least 6 square inches and not more than 55 square inches in cross-section area. Listed ducts for masonry fireplaces must be installed according to the terms of their listing and the manufacturer's instructions. Factory-built fireplaces require exterior air ducts that are a listed component of the fireplace and installed according to the fireplace manufacturer's instructions. **[Ref. R1006.1, R1006.4]**

Intakes for exterior air may be located on the exterior of the dwelling or in naturally ventilated attics or crawl spaces. The IRC does not permit the installation of combustion air intakes in garages, basements or mechanically ventilated spaces. Where combustion air openings are located inside the firebox, the exterior termination of the air intake cannot be higher than the firebox. Such an installation could result in combustion products being drawn to the outside through the air intake duct, creating a fire hazard. Exterior air intakes must be covered with a corrosion-resistant screen of ¼-inch mesh. **[Ref. R1006.2]**

MASONRY CHIMNEYS AND FIREPLACES

The prescriptive provisions of the IRC address structural support, approved materials, dimensions, and fire safety for masonry fireplaces. The code also prescribes construction details of masonry chimneys for proper drafting, weather protection, and safety from fire.

Design for masonry chimneys and fireplaces also must comply with other structural provisions of the code, including foundation requirements.

Footings

For masonry chimneys and fireplaces, the IRC requires concrete or solid masonry footings that are at least 12 inches thick and extend at least 6 inches beyond the face of the fireplace or foundation wall on all sides (Figure 11-2). [Ref. R1001.2, R1003.2]

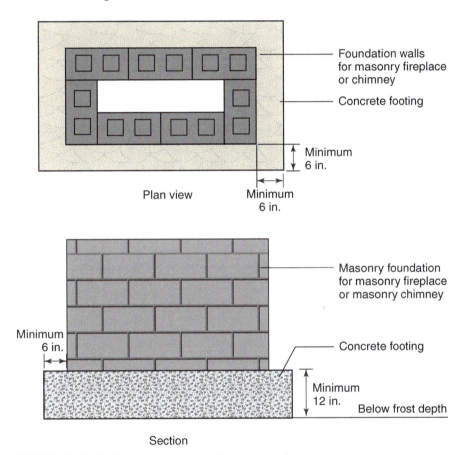

Plan view

Foundation walls for masonry fireplace or chimney

Concrete footing

Minimum 6 in.

Minimum 6 in.

Masonry foundation for masonry fireplace or masonry chimney

Concrete footing

Minimum 6 in.

Minimum 12 in.

Below frost depth

Section

FIGURE 11-2 Footings for masonry chimneys and fireplaces

Seismic requirements

Masonry without reinforcing is susceptible to damage from the ground motion of earthquakes. Chimneys in particular, because of their height and narrow width, often suffer severe damage or collapse in significant seismic events. Proper anchorage to the dwelling structure and the installation of reinforcing steel improve performance during earthquakes. The prescriptive seismic anchorage and reinforcing requirements in the IRC apply only to chimneys located in seismic design categories (SDCs) D_0, D_1 and D_2. No specific seismic requirements exist for chimneys in SDC A, B or C. Buildings in SDCs E and F require an engineered design in accordance with the *International Building Code* (IBC).

For masonry and concrete chimneys up to 40 inches wide with a single flue and located in SDC D_0, D_1 or D_2, the IRC requires four continuous vertical No. 4 reinforcing bars. Two additional vertical bars are required for each additional flue or each additional 40 inches of width or fraction thereof. Vertical reinforcing must be enclosed in ¼-inch horizontal ties spaced not more than every 18 inches vertically. Two horizontal ties are required wherever there are bends in the vertical bars. In addition, masonry or concrete chimneys require lateral support through prescribed anchorage at every floor, ceiling or roof more than 6 feet above grade (Figure 11-3). **[Ref. R1003.3, R1003.4, Table R1001.1]**

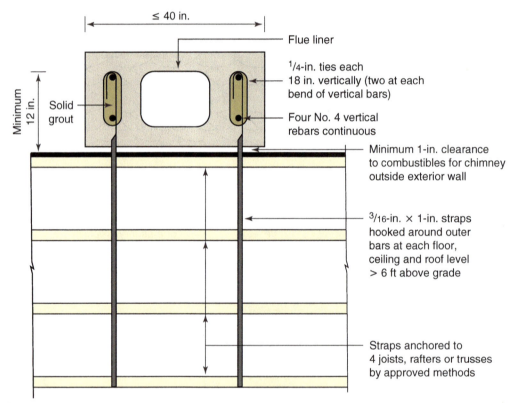

FIGURE 11-3 Masonry chimney seismic reinforcing and anchorage in SDCs D_0, D_1 and D_2

Masonry fireplace details

Solid masonry or concrete firebox walls with minimum 2-inch-thick fire brick lining require a total thickness of not less than 8 inches, including the lining. Prescribed dimensions for the firebox, throat, and smoke chamber facilitate the proper discharge of smoke and products of combustion through the chimney and intend to prevent downdrafts. Steel or other noncombustible lintels supporting masonry above the firebox require minimum 4-inch bearing support at each end. The code requires an operable steel or cast iron damper located not less than 8 inches above the firebox opening that can be closed when the fireplace is not in use (Figure 11-4). **[Ref. R1001.5 to R1001.8]**

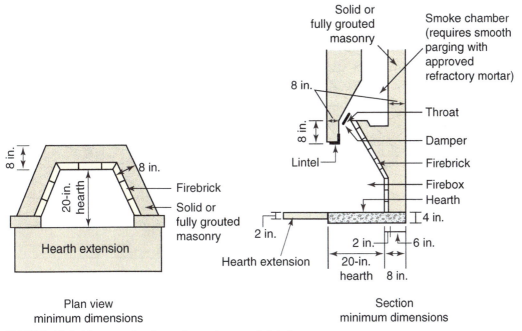

FIGURE 11-4 Masonry fireplace dimensions and details

Hearth and hearth extension

The hearth is the floor of the firebox and is constructed to withstand the intense heat of a wood fire. The hearth extension projects from the front of the firebox to provide a noncombustible area of protection against radiant heat and any embers, sparks or ash that may escape from the firebox. Hearths and hearth extensions must be constructed of concrete or masonry, supported by noncombustible materials, and reinforced to carry their own weight and all loads. The prescribed minimum thicknesses are 4 inches for the hearth and 2 inches for the hearth extension. When the firebox opening is raised at least 8 inches above the floor, the hazard of igniting underlying materials is reduced, and the code permits a hearth extension of ⅜-inch-thick brick, concrete, stone, tile or other approved noncombustible material supported by combustible construction. For fireplace openings with areas less than 6 square feet, hearth extensions must project at least 16 inches out from the face of the fireplace and at least 8 inches beyond each side of the opening. For larger openings, the dimensions are a minimum 20 inches in front of the face and a minimum 12 inches beyond each side of the fireplace opening (Figure 11-5). [Ref. R1001.9, R1001.10]

Clearance to combustibles and fireblocking

To prevent fires caused by conductive and radiant heat transfer, the IRC prescribes clearances to wood and other combustible materials in proximity to masonry fireplaces and chimneys. In general, wood floor, wall, ceiling, and roof framing requires a minimum 2-inch air clearance from the masonry, though this dimension is increased to 4 inches on the back of the fireplace. Combustible sheathing, siding, flooring, trim and gypsum board may abut the masonry fireplace and chimney sidewalls, provided that such combustible material maintains a distance of not less than 12 inches from the inside surface of the firebox and the inside surface of the flue (Figure 11-6). **[Ref. R1001.11, R1003.18]**

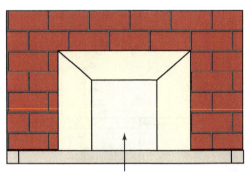
Fireplace opening ≥ 6 sq. ft.

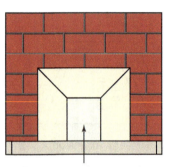
Fireplace opening < 6 sq. ft.

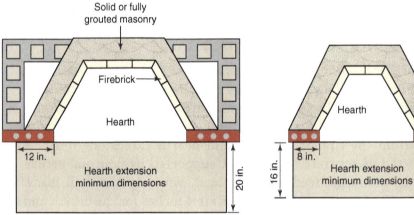

FIGURE 11-5 Hearth and hearth extensions for masonry fireplaces

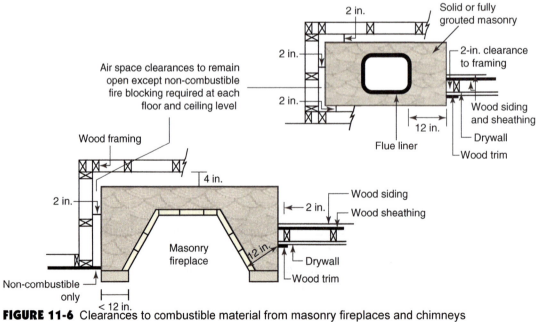

FIGURE 11-6 Clearances to combustible material from masonry fireplaces and chimneys

Combustible mantels and trim placed directly on the front of the fireplace are not restricted when located more than 12 inches from the fireplace opening. Such materials located more than 6 inches but not greater than 12 inches from the fireplace opening are permitted but must not project more than ⅛ inch from the face of the masonry for each inch of separation from the edge of the fireplace opening. The code does not permit combustible trim within 6 inches of the opening (Figure 11-7). **[Ref. R1001.11]**

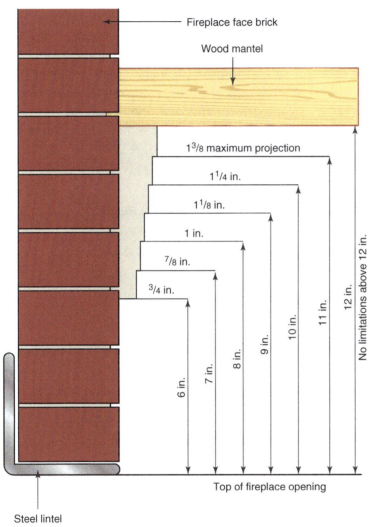

Fireplace face brick

Wood mantel

1³/8 maximum projection

1¹/4 in.

1¹/8 in.

1 in.

⁷/8 in.

³/4 in.

6 in.

7 in.

8 in.

9 in.

10 in.

11 in.

12 in.

No limitations above 12 in.

Top of fireplace opening

Steel lintel

FIGURE 11-7 Clearances to combustible mantels and trim on the face of masonry fireplaces

The IRC requires fireblocking at prescribed intervals to stop the spread of fire in concealed spaces. This rule applies to the air spaces created by clearance to combustible materials around fireplaces and chimneys. In the case of masonry chimneys and fireplaces, the fireblocking must be noncombustible and must be installed at each floor and ceiling line. **[Ref. R1001.12, R1003.19]**

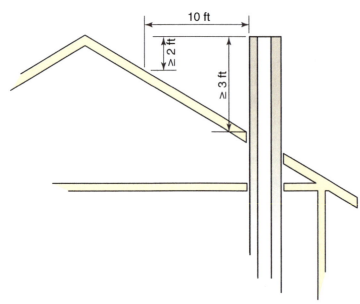

FIGURE 11-8 Masonry chimney termination

Chimney termination

To provide proper drafting, masonry chimneys must terminate at least 3 feet above the roof and at least 2 feet higher than any portion of a building within 10 feet (Figure 11-8). Flashing to weatherproof the chimney penetration at the roof must comply with the IRC roof covering and flashing requirements (see Chapter 7 of this publication). Where asphalt or wood shingles join the sides of a chimney, step flashing is required. For chimneys 30 inches wide or larger, the IRC also requires a cricket to direct water shed from the roof above around the sides of the chimney (Table 11-1 and Figure 11-9). **[Ref. R1003.9, R1003.20, R903.2, R905.2.8]**

TABLE 11-1 Cricket dimensions

Roof Slope	Height
12:12	½ of width
8:12	⅓ of width
6:12	¼ of width
4:12	⅙ of width
3:12	⅛ of width

[Ref. Table R1003.20]

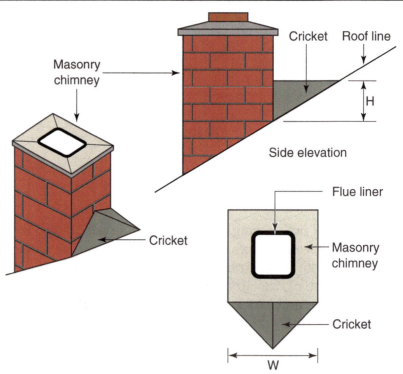

FIGURE 11-9 Cricket dimensions

Weather protection for masonry chimney terminations is accomplished with a required chimney cap and an optional rain cap. A chimney cap protects the top of the masonry surrounding the flue. The cap must be sloped to the outside and overhang the face of the masonry chimney to provide a drip edge. The prescribed caulking of the joint between the masonry and the flue serves as both a sealant and a bond break for any differential movement. Though not required, rain caps are installed a prescribed distance above a chimney flue termination to limit the amount of rain entering the flue while still providing adequate air flow for efficient venting of the products of combustion (Figure 11-10). **[Ref. R1003.9.1, R1003.9.3]**

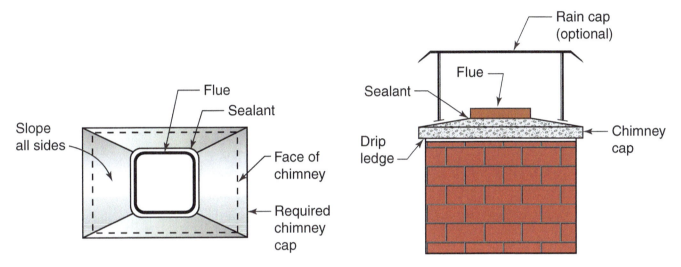

FIGURE 11-10 Masonry chimney cap and rain cap

MANUFACTURED CHIMNEYS AND FIREPLACES

Factory-built fireplaces and chimneys must be listed and labeled, tested in accordance with UL 127 and installed according to the conditions of the listing. The hearth extension dimensions prescribed for masonry fireplaces do not apply to manufactured fireplaces, which require hearth extensions to be installed in accordance with the listing of the fireplace. However, the IRC does require that hearth extensions be readily distinguishable from the surrounding floor area (Figure 11-11). **[Ref. R1004, R1005]**

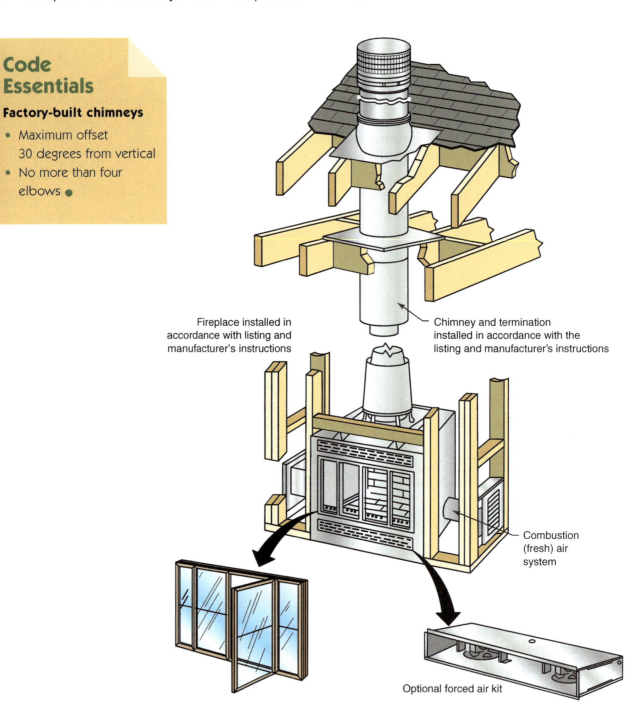

Fireplace installed in accordance with listing and manufacturer's instructions

Chimney and termination installed in accordance with the listing and manufacturer's instructions

Combustion (fresh) air system

Optional forced air kit

FIGURE 11-11 Factory-built fireplace and chimney

Building Utilities

12

Mechanical and Fuel Gas

The mechanical section of the *International Residential Code* (IRC) regulates the installation of permanently installed equipment and systems that control environmental conditions of a dwelling, such as comfort heating and cooling, including solid- or liquid-fuel appliances, duct systems and ventilation systems. The fuel-gas section of the code regulates the installation of natural gas and liquid petroleum gas (LPG or LP gas) piping systems and the associated gas-fired appliances, including provisions for combustion air and venting of combustion products. Installations of mechanical systems that are not covered in the IRC must comply with the applicable provisions of the *International Mechanical Code* (IMC) or *International Fuel Gas Code* (IFGC). Discussion in this chapter will focus on common heating, ventilating, and air conditioning (HVAC) systems, gas-fired appliances and gas piping systems.

APPLIANCES

Listing and labeling of appliances by qualified nationally recognized third party agencies, as mandated by the code, gives assurance that an appliance, when installed in accordance with the manufacturer's instructions, will function satisfactorily for the intended purpose and operate safely. The IRC requires the appliance to be installed and used in a manner consistent with the listing. For example, the listing may limit the use to a residential application in an indoor location. The required label is a factory applied nameplate identifying the manufacturer and the testing agency and providing other specified information. Labels for gas-fired appliances must indicate the hourly input rating in British thermal units per hour (Btu/h), the approved type of fuel (natural gas or LP gas), and the minimum clearances around the appliance (Figure 12-1). **[Ref. M1302, M1303, G2404.3]**

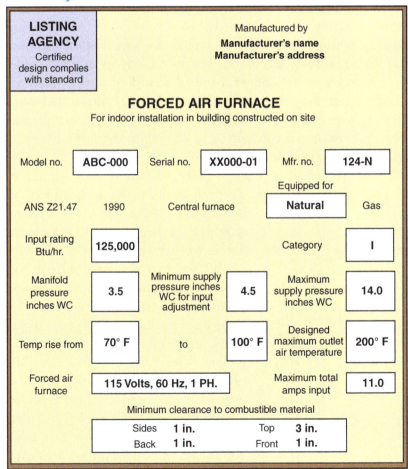

FIGURE 12-1 Appliance label

Appliance installation and location

Appliance installation must conform to the requirements of the IRC and to the conditions of the appliance listing. **[Ref. R102.4, M1307.1, M1401.1, G2406.1, G2408.1]**

Clearances

The appliance listing and manufacturer's installation instructions generally determine minimum clearance to combustibles and minimum air clearance around the appliance for proper operation. In some instances, clearance to combustibles may be reduced with the application of noncombustible insulating materials, provided that such is not prohibited by the appliance listing. **[Ref. M1306, G2408.5]**

Location limitations

With some exceptions, the IRC prohibits the installation of gas-fired appliances, such as furnaces and water heaters, in sleeping rooms, bathrooms, toilet rooms or storage closets, or in a space that opens only into such rooms or spaces. Such installation in small rooms with closed doors increases the risk of inadequate combustion air, improper operation, depleted oxygen levels, and exposure to carbon monoxide and other hazardous products of combustion if the appliance malfunctions. In addition, sleeping occupants are not alert to respond to developing hazardous conditions. Direct-vent appliances have sealed combustion chambers and draw all combustion air directly from the outside. Therefore, direct-vent appliances installed according to the manufacturer's instructions are permitted in these spaces. Certain vented room heaters, fireplaces and decorative appliances also may be installed in bedrooms, bathrooms or connecting spaces when the room contains the prescribed volume of combustion air.

With additional safeguards in place to isolate the appliance, the IRC does allow a gas-fired furnace, boiler or water heater to be installed in a room or space that opens only into a bedroom or bathroom. Access to such space must be through a solid, weather-stripped, self-closing door, and all combustion air must be taken directly from the outdoors. The code also prohibits the space from being used for any other purpose, such as storage (Figure 12-2). In addition, the code permits the installation of a gas-fired clothes dryer in a bathroom or toilet room provided that a transfer opening not less than 100 square inches communicates with another area of the dwelling unit. The opening cannot be to another bathroom or toilet room, a bedroom or a closet (Figure 12-3). **[Ref. G2406.2]**

Combustion air directly from outdoors

Non-direct vent, gas fired furnace, boiler or water heater

Self-closing solid door with weatherstripping

No storage or other use

Clothes closet

Bedroom

FIGURE 12-2 Gas appliance installed in a space that opens only into a bedroom

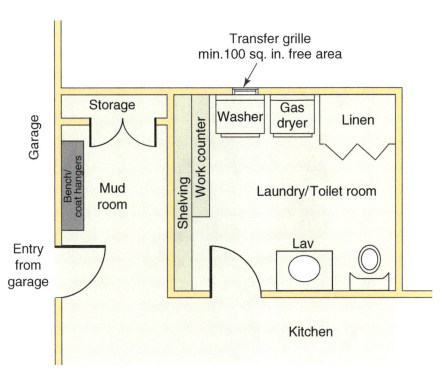

FIGURE 12-3 Gas clothes dryer in a toilet room

Appliances in garages

Vapors from gasoline and other flammable liquids often found in garages are heavier than air and may accumulate near the floor in sufficient concentration to ignite in the presence of a flame or spark. Accordingly, the IRC requires any ignition source of an appliance installed in a garage to be at least 18 inches above the floor unless the appliance is listed as flammable-vapor-ignition resistant. An *ignition source* is defined as a flame, spark, or surface capable of igniting flammable vapors or fumes, and includes appliance burners, burner igniters and electrical switching devices. **[Ref. M1307.3, G2408.2]**

Protection from impact

Accidental physical damage to an appliance or its fuel connection also creates a hazardous condition that may result in fire, explosion or improper appliance operation, and the IRC requires protection from impact by vehicles. Although appliance installations in garages are common and are most likely to be affected by close proximity to cars and trucks, outdoor locations, particularly those adjacent to driveways, also are vulnerable to vehicle impact and require protection. Protection may be achieved by installing bollards, curbs, or other approved barriers. Suspended appliances with sufficient clearance above the floor and appliances installed in an alcove out of the path of vehicle travel are not subject to impact and do not require additional barriers (Figure 12-4). The IRC also prohibits placing any stress on the connections of the fuel-gas piping system thereby reducing the possibility of damage to gas pipe fittings causing leaks or ignition of fuel gas. **[Ref. M1307.3.1, G2408.6]**

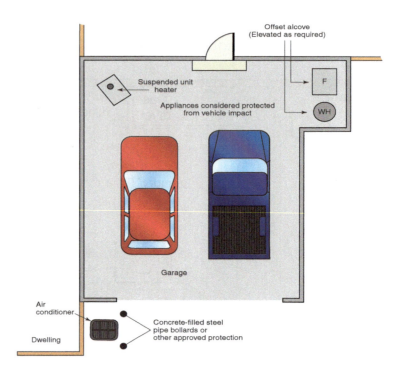

FIGURE 12-4 Appliance protection from vehicle impact

Exterior installation

Appliances installed outdoors must be listed for exterior locations (or provided with protection from the weather) and be supported on a level concrete slab or other approved material extending at least 3 inches above grade. When suspended, such installations require a minimum 6-inch clearance above grade. The clearance-to-grade requirements also apply to appliances installed in crawl spaces. **[Ref. M1305.1.3.1, M1401.4, G2408.4]**

Access to appliances

In addition to the appliance manufacturer's installation instructions, the IRC provides for adequate access and clearance to facilitate the service, repair, and replacement of appliances. All appliances require a minimum 30-inch by 30-inch working space in front of the controls. Access doors and passageways to appliances must be at least 24 inches wide and large enough to remove the largest appliance. **[Ref. M1305]**

Appliances in attics

As with other locations, appliances installed in attics require a sufficient access opening to remove the largest appliance. Minimum opening size is 30 inches by 20 inches. Because of the difficulty of accessing and servicing appliances in attics unless they are located adjacent to the access opening, the code places restrictions on the distance to the appliance from the access opening and provides for a solid surface passageway not less than 24 inches wide. The length of the passageway is limited to not more than 20 feet unless there is a clear path at least 22 inches wide by 6 feet high, in which case the appliance may be located as much as 50 feet from the access opening. A solid platform is still required to satisfy the minimum 30-inch by 30-inch working space at the service side of the appliance. The IRC further requires a light fixture and receptacle outlet at the appliance location. The light must be controlled by a switch located at the access opening (Figure 12-5). Similar provisions apply to appliances installed under floors (crawl spaces). [Ref. M1305.1.3]

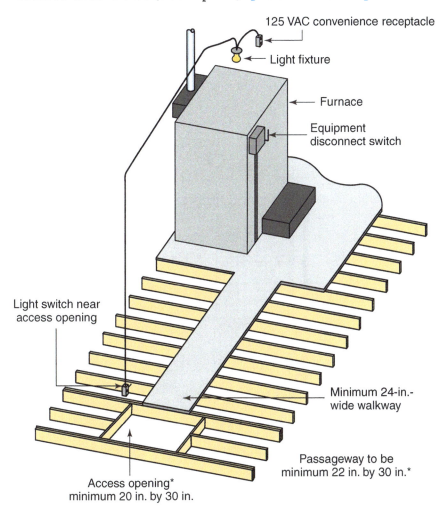

125 VAC convenience receptacle

Light fixture

Furnace

Equipment disconnect switch

Light switch near access opening

Minimum 24-in.-wide walkway

Passageway to be minimum 22 in. by 30 in.*

Access opening* minimum 20 in. by 30 in.

*Large enough to allow removal of largest piece of equipment

FIGURE 12-5 Attic installation requirements

Condensate

The IRC regulates the proper disposal of condensate that results from either cooling or combustion processes. Cooling coils and evaporators installed in forced air furnaces as part of the air-conditioning system generate condensate. High-efficiency category IV furnaces have low- temperature flue gases that also produce condensate in the vent. In many cases, both conditions exist in the same appliance. In general, condensate must drain to an approved location such as a floor drain. Where the appliance is installed in an upper story or an attic, where water leakage will cause damage to building components such as the drywall ceiling of the living space below, additional preventive measures are required.

The most common method to prevent water damage to construction materials because of a stoppage in the primary drain is to install an auxiliary drain pan below the appliance. In addition to prescribing the pan dimensions and materials, the IRC requires discharge to a conspicuous location to alert occupants of a problem. As an alternative to the auxiliary drain pan, the code permits installation of a secondary ¾-inch drain line from the appliance's integral drain pan discharging to a conspicuous location. Acceptable alternatives to the auxiliary pan or secondary drain provide for automatic shut-down of the appliance when a stoppage in the drain occurs (Figure 12-6). [Ref. M1411.3, M1411.5, G2404.10]

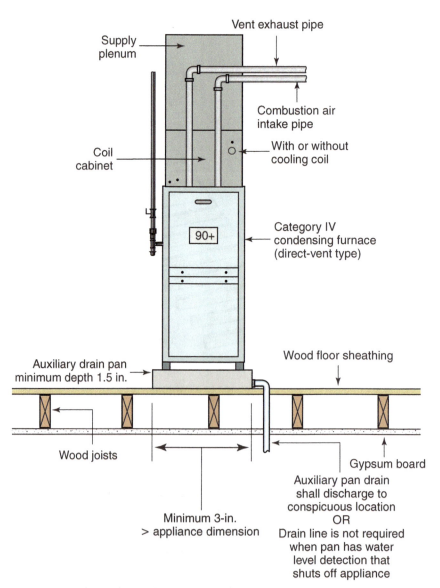

Vent exhaust pipe

Supply plenum

Combustion air intake pipe

Coil cabinet

With or without cooling coil

90+

Category IV condensing furnace (direct-vent type)

Wood floor sheathing

Auxiliary drain pan minimum depth 1.5 in.

Wood joists

Gypsum board

Auxiliary pan drain shall discharge to conspicuous location
OR
Drain line is not required when pan has water level detection that shuts off appliance

Minimum 3-in. > appliance dimension

FIGURE 12-6 Auxiliary drain pan for condensate

EXHAUST SYSTEMS

Mechanical exhaust systems discharge airborne contaminants and moisture to the outside atmosphere. For other than ventilation-type attic fans discharging to the attic, the IRC prohibits exhaust systems to terminate in an attic, soffit, ridge vent or crawl space to prevent damage to the structure from moisture. [Ref. M1501]

Clothes dryer exhaust systems

Exhaust systems for electric and gas clothes dryers must be installed in accordance with the appliance listing and the manufacturer's instructions. Dryer exhaust ducts convey moisture and, in the case of gas dryers, combustion products to the outdoors. Because dryers discharge combustible lint, the code prescribes measures to prevent lint buildup, thereby reducing the hazard of a fire. In addition to conforming to the manufacturer's instructions, ducts must be 4 inches nominal in diameter and constructed of smooth, rigid metal at least No. 28 gage (0.0157 inches thick). Joints must be assembled in the direction of air flow, and fasteners cannot protrude more than $1/8$ inch into the inside of the duct.. Maximum duct length is determined by the code-specified length of 35 feet less the tabular reductions for fittings based on the degree and radius of the bend or maximum length is determined by the dryer manufacturer's instructions. The IRC also permits the installation of a dryer exhaust duct power ventilator installed in accordance with the manufacturer's instructions. The code limits connectors between the dryer and the rigid exhaust duct to a single piece of approved listed and labeled transition duct not more than 8 feet long. The connector cannot be concealed.

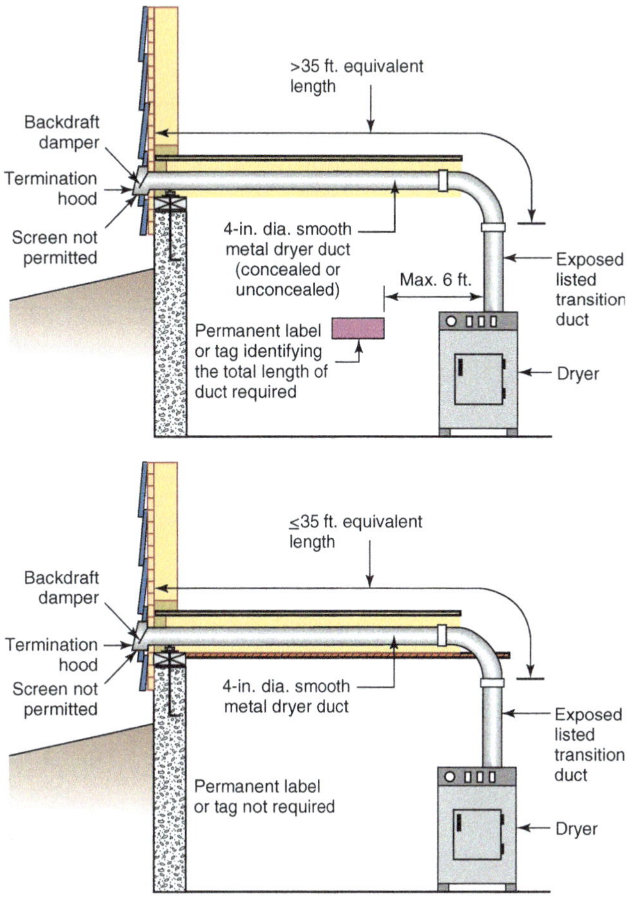

To prevent discharged air from reentering the building, dryer exhaust ducts must terminate outside at least 3 feet from openings such as windows, doors, or ventilation intake locations. The IRC requires a backdraft damper and prohibits the installation of screens at the termination point (Figures 12-7 through 12-10). [Ref. M1502, M1506.2, G2439]

FIGURE 12-7 Clothes dryer exhaust system

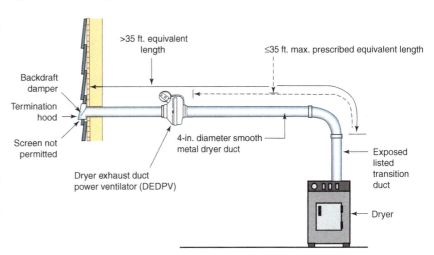

FIGURE 12-8 Dryer duct power ventilator installed to increase duct length

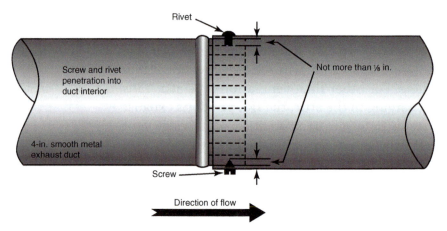

FIGURE 12-9 Dryer duct joints must be inserted in direction of flow and fasteners cannot penetrate more than $1/_8$ inch.

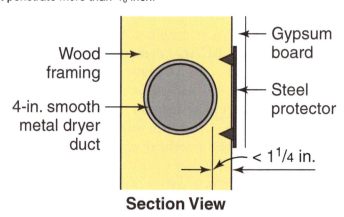

Section View

FIGURE 12-10 Protection of concealed dryer duct from fastener penetration

Whole-house mechanical ventilation system

With an increased awareness of and emphasis on energy conservation, and in keeping with the latest provisions of the *International Energy Conservation Code* (IECC), new house construction increasingly relies on insulation, air barriers and sealants to provide a tighter thermal envelope, which significantly reduces natural infiltration of outside air. Although tight construction is beneficial in reducing air leakage and conserving energy, closed-house conditions in the heating or cooling season may lead to inadequate fresh air and poor indoor air quality. A whole-house ventilation system simply exchanges indoor air for outdoor air and increasingly is necessary to provide adequate fresh air to dwelling units. The minimum air-flow rates prescribed in the IRC mechanical provisions are based on the area of the dwelling and the number of bedrooms. When a whole-house ventilation system is installed, the code permits a combination of supply and exhaust fans for achieving adequate ventilation and does not require a separate energy recovery ventilation system. See Chapters 10 and 15 for additional discussion of ventilation and energy conservation. [Ref. M1505]

Kitchen range hoods

Domestic open-top broiler units require an overhead metal exhaust hood unless the unit is equipped with a listed integral system. Exhaust hoods are not required for other domestic kitchen cooking appliances. However, when installed, range hoods must comply with the manufacturer's instructions and the code requirements. For other than downdraft systems and listed ductless (recirculating) hoods, the IRC requires a single-wall, smooth, airtight duct of galvanized steel, stainless steel or copper terminating outdoors with a backdraft damper. In addition, to prevent negative pressure and the accompanying adverse effects on mechanical appliances and systems within the dwelling, makeup air is required for high-velocity fans exceeding an exhaust capacity of 400 cfm, unless all fuel-fired appliances are direct vent or mechanical draft vented. (Table 12-1 and Figure 12-11). **[Ref. M1503]**

TABLE 12-1 Minimum required local exhaust rates

Area to be exhausted	Exhaust rates
Kitchens	100 cfm intermittent or 25 cfm continuous
Bathrooms—toilet rooms	Mechanical exhaust capacity of 50 cfm intermittent or 20 cfm continuous

[Ref. Table M1505.4.4]

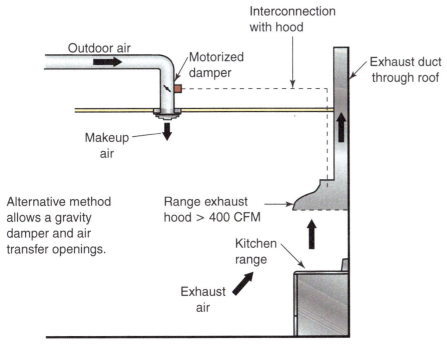

FIGURE 12-11 Required makeup air for kitchen exhaust hoods exceeding 400 cfm unless all fuel-fired appliances are direct vent or mechanical draft vented

Bathrooms and toilet rooms with mechanical ventilation

Unless there is an openable window, bathrooms and toilet rooms require mechanical ventilation to exhaust air directly to the outdoors. The IRC prohibits the recirculating of the exhaust air within the dwelling. Ventilation rates must comply with Table 12-1. **[Ref. M1505.2, M1505.4.4, R303.3]**

DUCT SYSTEMS

In order to effectively and safely circulate environmental air, duct systems must be fabricated of approved materials that meet the temperature, flame-spread, and smoke-developed ratings of the code. Factory-made ducts must be listed and labeled as complying with applicable standards. The IRC is concerned primarily with the burning characteristics of duct materials, including insulation, and the hazards related to the spread of smoke in the case of a fire. In addition to these fire-safety requirements, the IRC prescribes airtight joints and seams and installation to adequately ensure the structural integrity of the duct system. **[Ref. M1601]**

The IRC mechanical provisions permit isolated interior stud-wall cavities and panned solid-joist spaces to serve as return air ducts, provided that they are not part of a fire resistance rated assembly, do not convey return air from more than one floor level, and are properly fire blocked. However, because it is difficult to seal framing cavities used as air plenums and such spaces are considered inefficient for moving air, the energy conservation provisions of the code do not allow building framing cavities to be used as return air plenums. **[Ref. M1601.1.1]**

RETURN AIR

Return air is typically the air removed from an approved conditioned space and recirculated through the HVAC system. The routine operation of a dwelling may cause available return air to be lost through exhaust systems, appliance vents, and fireplace chimneys. Recognizing this, the IRC permits return air to be diluted or supplemented with fresh outdoor air through outdoor air inlets that must be screened.

The IRC restricts the sources of return air to prevent the circulation of unpleasant or noxious odors or other contaminants and to prevent negative pressures in confined spaces or appliance locations. **[Ref. M1602, G2442.4]**

COMBUSTION AIR

Fuel-burning appliances require a supply of air for fuel combustion, draft hood dilution, and ventilation of the space in which the appliance is installed. Combustion air is important not only for proper operation of the appliance but for protecting occupants against the hazards of oxygen depletion and buildup of harmful combustion gases. The IRC references the combustion air requirements of the manufacturer for solid fuel-burning appliances and NFPA 31 for oil-fired appliances. Discussion here focuses on the specific combustion air requirements for gas-fired appliances. **[Ref. M1701, G2407.1]**

Calculating the net free area of vents or grilles

When using louvers or grilles to satisfy combustion air requirements, the amount of net free area is as specified by the manufacturer, if known. If not known, the amount can be calculated as 75 percent of the gross area for metal louvers and 25 percent of the gross area for wood louvers (Figure 12-12). **[Ref. G2407.10]**

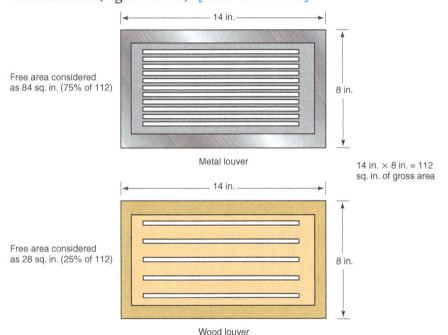

14 in.

Free area considered as 84 sq. in. (75% of 112)

Metal louver

8 in.

14 in. × 8 in. = 112 sq. in. of gross area

14 in.

Free area considered as 28 sq. in. (25% of 112)

Wood louver

8 in.

FIGURE 12-12 Determining louver net free area when not specified by the manufacturer

Combustion air from inside the building

Combustion air for gas-fired appliances may be obtained from an indoor space having a volume of at least 50 cubic feet per 1000 Btu/h input rating of all appliances being served within the space. The IRC also allows drawing combustion air from adjacent rooms through two permanent openings. One opening must be within 12 inches of the ceiling and one must be within 12 inches of the floor. The code requires each opening to have a free area of not less than 100 square inches and at least 1 square inch per 1,000 Btu/h input rating of all appliances installed within the space (Figure 12-13). **[Ref. G2407.5.1, G2407.5.3]**

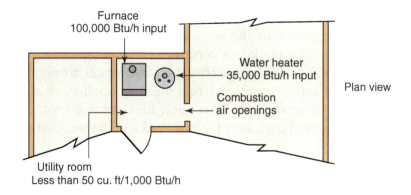

Example: Calculate indoor combustion air volume:

$$\frac{135,000}{1,000} \times 50 = \begin{array}{l} 6,750 \text{ cu. ft} \\ \text{volume required} \end{array}$$

Example: Determine net free area for each combustion air opening
Total appliance input -135,000 Btu/h

$$\frac{135,000}{1,000} = \begin{array}{l} 135 \text{ sq in. net free} \\ \text{area per opening} \end{array}$$

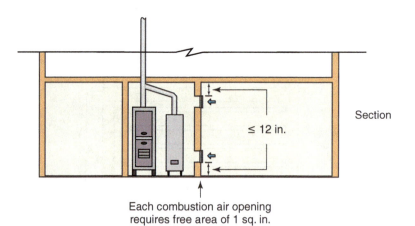

Each combustion air opening
requires free area of 1 sq. in.
per 1,000 Btu/h appliance input

FIGURE 12-13 Indoor combustion air from spaces on same story

Combustion air from outdoors

For gas-fired appliances, the IRC prescribes two methods for obtaining combustion air from the outdoors or from a space freely communicating with the outdoors, such as a ventilated attic. **[Ref. G2407.6]**

Outdoor combustion air obtained through two openings or ducts

In the first method for obtaining outdoor combustion air, two openings or ducts are required, one within 12 inches of the ceiling and one within 12 inches of the floor. Each vertical duct or direct opening to the outdoors requires a free area of at least 1 square inch per

4,000 Btu/h of total input rating. Horizontal ducts require a larger cross-sectional free area of at least 1 square inch per 2,000 Btu/h of total input rating. For combustion air ducts that terminate in an attic, the IRC requires the termination point to be not less than 6 inches above the top of the ceiling joists and insulation, and does not permit screens on the termination inlet (Figure 12-14). **[Ref. G2407.6.1]**

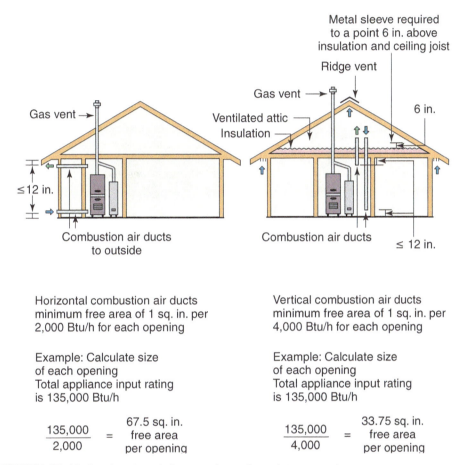

FIGURE 12-14 Combustion air from outdoors through two openings

Outdoor combustion air obtained through a single opening or duct

The IRC also permits combustion air for gas-fired appliances to be obtained through a single opening located within 12 inches of the ceiling when the size is increased to meet three criteria. The free area of the opening must be at least 1 square inch per 3,000 Btu/h of the total appliance input rating and must be at least the sum of the areas of all vent connectors in the space. The code also prescribes minimum clearances around the appliances for free circulation of air (Figure 12-15). **[Ref. G2407.6.2]**

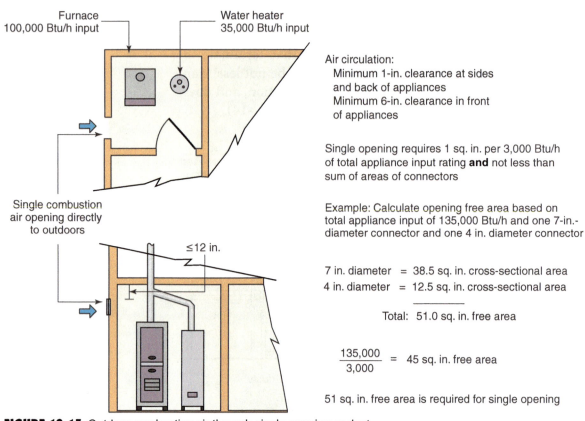

FIGURE 12-15 Outdoor combustion air through single opening or duct

---

VENTS

For gas-fired appliances, the code prescribes the methods for venting combustion products to the outside atmosphere. For other than approved plastic piping used for venting category IV appliances, vents must be listed and labeled for use with the type of appliance.

Vent installation

The code addresses two concerns of gas vent installation: separation from combustible materials to prevent ignition and protection of the vent from physical damage. The vent listing and manufacturer's instructions determine minimum clearances to combustibles. For vents passing through insulated attics, a 26-gauge sheet metal insulation shield that terminates at least 2 inches above the insulation is required. Where vents are installed in concealed locations through holes or notches in framing and are less than 1½ inches from the edge of the framing, the vents require protection from fastener penetration by installing a fastener shield plate of 16-gauge steel (Figure 12-16). Vent terminals for sidewall vented appliances, such as direct-vent fireplaces, heaters, water heaters, furnaces and boilers, must be located such that doors cannot swing within 12 inches of the vent terminal. This measure protects the vent from physical damage and assures adequate clearance for proper operation of the vent system (Figure 12-17). **[Ref. G2426]**

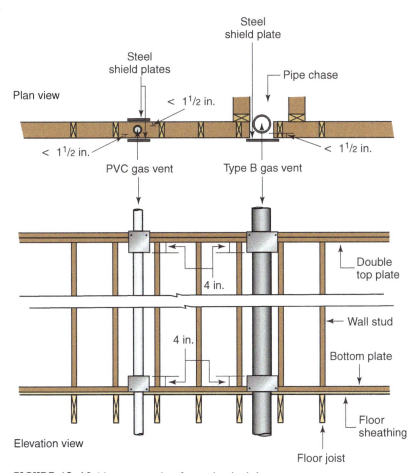

Steel
shield plate

Steel
shield plates

Pipe chase

Plan view

< 1¹/₂ in.

< 1¹/₂ in.

< 1¹/₂ in.

PVC gas vent

Type B gas vent

4 in.

Double
top plate

4 in.

Wall stud

Bottom plate

Floor
sheathing

Elevation view

Floor joist

FIGURE 12-16 Vent protection from physical damage

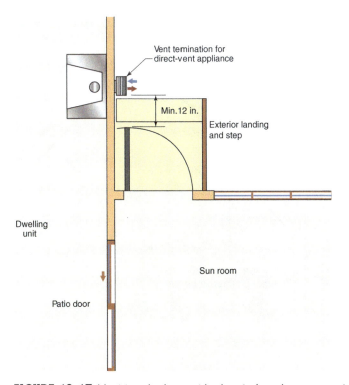

Vent temination for
direct-vent appliance

Min.12 in.

Exterior landing
and step

Dwelling
unit

Sun room

Patio door

FIGURE 12-17 Vent terminals must be located so doors cannot
swing within 12 inches to protect against physical damage.

Gas-vent roof termination

The required termination height for gas vents not more than 12 inches in size and located at least 8 feet from a vertical wall is based on the roof pitch (Table 12-2 and Figure 12-18). Gas vents larger than 12 inches or less than 8 feet from a vertical wall must terminate at least 2 feet above the roof or any portion of a building within 10 feet horizontally. [Ref. G2427.6.3]

TABLE 12-2 Gas vent termination height for listed caps 12 inches and smaller located at least 8 feet from a vertical wall

Roof slope	Minimum height (ft) from roof to lowest discharge opening
Flat to 6/12	1.0
Over 6/12 to 7/12	1.25
Over 7/12 to 8/12	1.5
Over 8/12 to 9/12	2.0
Over 9/12 to 10/12	2.5
Over 10/12 to 11/12	3.25
Over 11/12 to 12/12	4.0
Over 12/12 to 14/12	5.0
Over 14/12 to 16/12	6.0
Over 16/12 to 18/12	7.0
Over 18/12 to 20/12	7.5
Over 20/12 to 21/12	8.0

[Ref. Figure G2427.6.3]

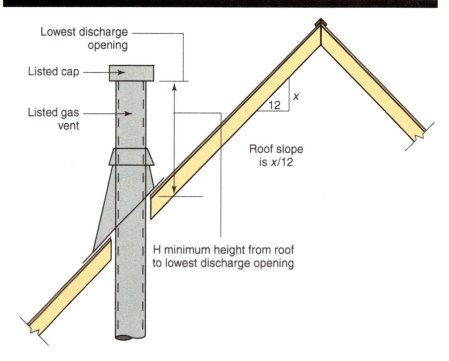

Lowest discharge opening

Listed cap

Listed gas vent

12   x

Roof slope is $x/12$

H minimum height from roof to lowest discharge opening

FIGURE 12-18 Gas-vent roof termination

Direct-vent appliance vent termination

Direct-vent appliances draw all combustion air directly from outside. Direct-vent appliances often produce lower-temperature flue gases that may be vented through the exterior wall and terminate near the combustion air intake location. Clearances from vent terminations to building openings are less than would otherwise be required for non-direct-vent appliances (Table 12-3). However, clearances must also comply with the listing of the appliance and the manufacturer's instructions. [Ref. G2427.8]

TABLE 12-3 Gas vent termination for direct-vent appliances

Appliance btu/h input rating		Min. clearance to air openings into building (in.)	Min. clearance above grade (in.)
Over	Not over		
–	10,000	6	12
10,000	50,000	9	12
50,000	150,000	12	12

[Ref. Table G2427.8]

FUEL-GAS PIPING

The IRC provides for the design, materials and safe installation of fuel-gas piping to serve gas-fired appliances.

Pipe sizing

Gas piping is sized to supply adequate volume to meet the demand of the connected appliances. Pipe size is determined based on a number of variables, including appliance input ratings, type and specific gravity of the gas, pipe material, length of pipe, inlet pressure and pressure drop. The code includes pipe sizing tables and sizing equations but permits other approved methods for sizing gas pipe, including manufacturer's instructions and engineered methods. [Ref. G2413]

Piping materials

Approved gas piping materials include Schedule 40 steel, Schedule 10 steel, approved seamless metallic tubing if gas used is not corrosive to the material and corrugated stainless steel tubing (CSST). Approved plastic pipe, tubing and fittings are permitted in exterior underground installations. Fittings and thread joint compounds when used must be compatible with the piping material and gas and approved for the specific use. [Ref. G2414]

Piping system prohibited locations

The IRC does not permit the installation of gas piping within an air duct, clothes chute, chimney, or gas vent or through any townhouse unit other than the unit being served. In addition, gas piping is not permitted to penetrate foundation walls below grade and must enter and exit a building at a point above grade. [Ref. G2415.3, G2415.6]

Other installation requirements

Drilling and notching of wood floor, wall, and roof framing is limited to locations and dimensions as specified in Chapter 6 of this publication. Concealed piping installed through holes or notches in framing members or installed parallel and alongside framing members, and less than 1½ inches from the nearest edge of the member must be protected by steel shield plates. Schedule 40 black or galvanized steel gas piping resists penetration and does not require such protection. CSST gas tubing requires protection in accordance with the code and the manufacturer's installation instructions. [Ref. M1308.2, G2415.7]

Above-ground piping outdoors requires a clearance of 3½ inches above ground and above roof surfaces. Protection from corrosion, such as painting or galvanizing, is required for exposed exterior ferrous metal piping. Underground piping, which is often polyethylene plastic, must be approved for the location. Galvanizing is not considered adequate protection from corrosion for underground steel pipe, which requires wrapping with approved material. Underground piping must be buried at least 12 inches deep. The IRC requires inspection and pressure testing of all fuel-gas piping systems before they are concealed or put into service. [Ref. G2415.9, G2415.11, G2417]

Appliance connections

Rigid metallic pipe and fittings, CSST, and listed and labeled appliance connectors are approved for appliance connection to the gas piping system. Connectors are not allowed to pass through walls, floors, partitions, ceilings, or appliance housings (other than connectors to fireplace inserts with proper grommets in accordance with the manufacturer's instructions). For other than rigid metallic pipe, connector length is limited to no more than 6 feet (Figure 12-19). [Ref. G2422]

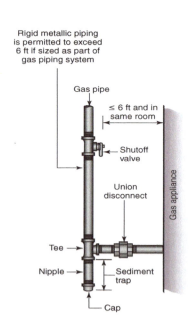

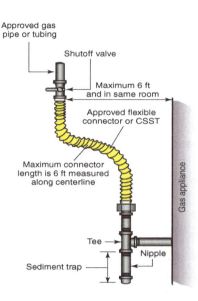

FIGURE 12-19 Gas appliance connection, shutoff valve and sediment trap

Shutoff valve

To facilitate service and replacement, each appliance requires an accessible shutoff valve located upstream of the connector in the same room and within 6 feet of the appliance. The code permits shutoff valves located as much as 50 feet from the appliance when installed at a manifold and clearly identified (Figures 12-19 and 12-20). [Ref. G2420.5]

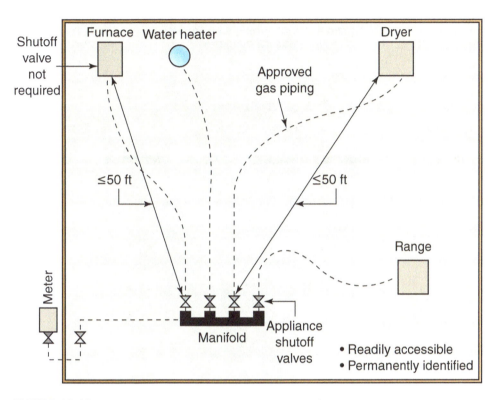

FIGURE 12-20 Appliance shutoff valves located at a manifold

Sediment trap

A sediment trap is generally required downstream of the shutoff valve and adjacent to the inlet of equipment. The IRC does not equire sediment traps for illuminating appliances, ranges, clothes dryers and outdoor grills (Figure 12-19). [Ref. G2419.4]

Piping support

Adequate piping support is necessary to prevent stresses on the pipe, fittings and connections. See Table 12-4 for maximum spacing of supports for gas piping. CSST supports must follow the manufacturer's instructions. **[Ref. G2424]**

TABLE 12-4 Support of fuel-gas piping

Steel pipe, nominal size (in.)	Spacing of supports (ft)	Nominal size of tubing smooth-wall (in., outside diameter)	Spacing of supports (ft)
½	6	½	4
¾ or 1	8	⅝ or ¾	6
1¼ or larger (horizontal)	10	⅞ or 1 (horizontal)	8
1¼ or larger (vertical)	Every floor level	1 or larger (vertical)	Every floor level
[Ref. Table G2424.1]			

CHAPTER
13
Plumbing

This chapter covers plumbing system design and installations typical of dwelling construction (Figure 13-1). Methods and materials outside the scope of the IRC must comply with the *International Plumbing Code* (IPC).

FIGURE 13-1 PEX water distribution tubing and PVC drain, waste, vent (DWV) piping

PIPING

Although the IRC recognizes many types of approved piping materials, discussion here will focus on those materials most commonly encountered in construction of one- and two-family dwellings and townhouses. Materials must be third-party certified as meeting the applicable IRC referenced standards. In order to perform as intended and to prevent damage or contamination, piping requires adequate support and protection from physical damage. **[Ref. P2609]**

Protection from damage

Concealed piping installed through holes or notches in studs, joists, or rafters and less than $1^1/_4$ inches from the nearest edge of the framing member requires protection from fastener penetration by shield plates. Protective shield plates must be at least 16-gauge steel and cover the area where the pipe passes through the member. Shield plates must extend at least 2 inches above bottom plates and below top plates of wall framing. Cast iron and galvanized steel pipe are sufficiently resistant to penetration by nails or screws and do not require shield plate protection (Figure 13-2). Limitations on boring and notching of structural members are covered in Chapter 6 of this publication. Pipes passing through foundation walls require a pipe sleeve. **[Ref. P2603]**

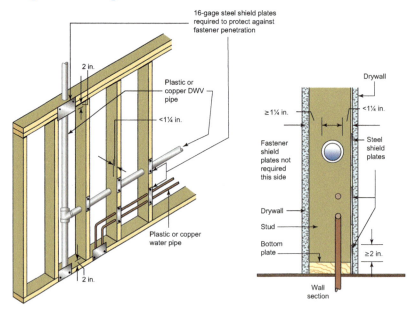

FIGURE 13-2 Physical protection of concealed piping

Water, drain and sewer piping must be protected from freezing. The IRC requires water service pipe to be buried at least 12 inches deep and at least 6 inches below the frost line. Because sewer lines are less susceptible to freezing, the code does not specifically require burial below the frost line. Minimum building sewer depth is determined by the building official and stipulated in the adopting ordinance. **[Ref. P2603.5]**

Piping support

Proper support of piping is important in maintaining alignment and slope, preventing sagging, and allowing for expansion and contraction. Piping installed underground requires continuous support on suitable bedding materials. Backfill over pipe must be free of debris, rocks, concrete and frozen material. For above-ground installations, the IRC prescribes the maximum support spacing for horizontal and vertical piping based on the pipe material. A mid-story guide is required for vertical plastic piping 2 inches and smaller in diameter (Table 13-1). [Ref. P2604, P2605, Table P2605.1]

TABLE 13-1 Piping support

Piping material	Max. horizontal spacing (ft)	Max. vertical spacing (ft)
ABS	4	10
Cast-iron,<10 ft lengths	5	15
Cast-iron, 10 ft lengths	10	15
Copper or copper alloy	12	10
PEX, ≤ 1 in. dia.	2.67	10
PVC	4	10

[Ref. excerpt of Table P2605.1]

ABS = acrylonitrile butadiene styrene, PEX = cross-linked polyethylene, PVC = polyvinyl chloride.

Testing of piping systems

The IRC requires pressure testing of sewer, drain-waste-vent (DWV) and water piping systems to ensure that there are no leaks. The building sewer is water tested with not less than a 10-foot head of water. The DWV system requires a water test with a 5-foot head of water or an air test maintaining 5 pounds per square inch (psi) of pressure. However, for other than PEX water piping, the code does not permit testing of plastic piping with air. The water supply system must be tested with a water pressure not less than the working pressure of the system. As an alternative, copper and PEX water piping may be tested with air at a minimum pressure of 50 psi. All tests must be maintained for 15 minutes without leakage. [Ref. P2503]

WATER SYSTEM

An approved potable-water supply is required for each dwelling unit, and the IRC prescribes methods for protecting and maintaining the system to deliver water that is safe for consumption by the occupants. The code also regulates design for adequate pressure and volume, suitable materials and fittings, and the location of shutoff valves.

Code Essentials

DWV water test

- Piping system is filled with water to detect leaks.
- May test in stages or the entire system at one time.
- Requires 5 feet of water-filled vertical pipe above all piping being tested.
- Five-foot head pressure not required on top 5 feet of DWV piping (usually vent) in building.
- Five-foot column of water applies a pressure of 2.17 psi.
- In multistory buildings, lowest parts of the system tested may be under significantly higher pressure. ●

Water service

Water service pipe is permitted in the same trench with a building sewer if the sewer pipe is listed for underground use within a building. For example, cast-iron or schedule 40 PVC DWV pipe is approved for this installation. For other types of building sewer pipe, such as polyethylene SDR-PR or PVC sewer and drain DR-PS pipe, the risk of contamination of the water supply is deemed greater, and the code requires separation. In these cases, the water service must have at least 5 feet of horizontal separation or be installed on a ledge at least 12 inches above and to one side of the highest point of the building sewer. [Ref. P2906.4.1]

Water supply system design criteria

The water system must be designed and installed to deliver adequate water volume and pressure for plumbing fixtures to operate efficiently and properly. The maximum static pressure for the water service at the building entrance is 80 psi. Even if adequate pressure and volume are provided with a smaller pipe, the minimum size for water service pipe is ¾ inch. Water supply fixture unit values, developed length of piping, and water pressure determine pipe size for the distribution system. The IRC also limits flow rates and consumption for plumbing fixtures to conserve water. [Ref. P2903]

Valves

A main shutoff valve is required for every dwelling unit near the entrance of the water service. The valve must be of a full open type with a provision for drainage of the water distribution system. All valves must be accessible. Other required shutoff valve locations are shown in Table 13-2. [Ref. P2903.9]

TABLE 13-2 Shutoff valve locations

Appliance or fixture	Valve location	Valve type
Water heater	Cold-water supply pipe at water heater	Full open
Lavatory	Each fixture supply pipe	Any approved type
Sink		
Water closet		
Bidet		
Bathtub	Not required	
Shower		
Hose bibb subject to freezing	Inside building*	Stop-and-waste type*

*Not required for frostproof hose bibb extending into heated building.

Dwelling unit fire sprinkler systems

The IRC plumbing provisions include a simple, prescriptive approach to the design of dwelling automatic fire sprinkler systems as an equivalent alternative to NFPA 13D systems. A dwelling fire sprinkler system requires less water when compared to NFPA 13 and 13R systems. The water supply requirements may be satisfied by a connection to a domestic water supply, a water well, an elevated storage tank, a pressure tank, or a stored water source with an automatically operated pump.

The code permits either a multipurpose sprinkler system, where domestic water is supplied to both sprinklers and plumbing fixtures, or a stand-alone sprinkler system. The code requires a rough-in inspection before piping is covered and a final inspection of the dwelling sprinkler system. Dwelling unit fire sprinkler systems are covered further in Chapter 9 of this publication. [Ref. P2904]

Water supply protection

The IRC requirements intend to protect the potable-water supply from contamination. Hose connections, boilers, heat exchangers, and lawn irrigation systems require listed back-flow prevention devices suitable for the application. The simplest and most effective means of preventing contamination from drain water and the associated bacteria is through the use of an air gap. Sinks, lavatories and bathtubs are examples of plumbing fixtures utilizing an air gap, which is the distance between the water outlet and the flood rim level of the fixture. The minimum air gap varies according to fixture type and application (Table 13-3 and Figure 13-3). A backflow preventer is not required between a stand-alone dwelling fire sprinkler system and the water distribution system. [Ref. P2902, Table P2902.3.1, P2904.1]

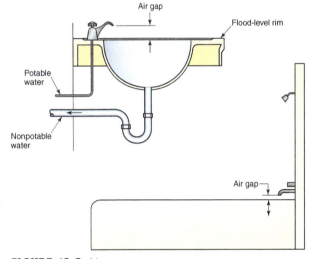

FIGURE 13-3 Air gaps

TABLE 13-3 Minimum air gaps

Fixture	Minimum air gap	
	Away from a wall (in.)	**Close to a wall (in.)**
Effective openings greater than 1 inch	Two times the diameter of the effective opening	Three times the diameter of the effective opening
Lavatories and other fixtures with effective opening not greater than ½ inch in diameter	1	1.5
Over-rim bath fillers and other fixtures with effective openings not greater than 1 inch in diameter	2	3
Sinks, laundry trays, gooseneck back faucets and other fixtures with effective openings not greater than ¾ inch in diameter	1.5	2.5
[Ref. Table P2902.3.1]		

SANITARY DRAINAGE

Proper operation of the sanitary drainage system depends on adequate pipe size of approved materials, appropriate slope and support, transition fittings suitable for the location, adequate cleanouts and proper venting.

Connections and fittings

The code requires specific approved methods for joining and connecting drainage piping to provide a gas-tight system with optimal flow. Cast-iron pipe is typically connected with compression gasket joints or mechanical couplings. PVC DWV pipe joints are solvent cemented. These two different piping materials may be joined together with approved mechanical couplings. The code prohibits a reduction in size in the direction of flow. Fittings for change in direction must be suitable for the installation to maintain acceptable flow and reduce the possibility of stoppage. For example, sanitary tees are not permitted for vertical-to-horizontal or horizontal-to-horizontal transitions. Such changes in direction are commonly accomplished with wye, combination wye and eighth bend, or long sweep fittings (Figure 13-4). **[Ref. P3003, P3005.1]**

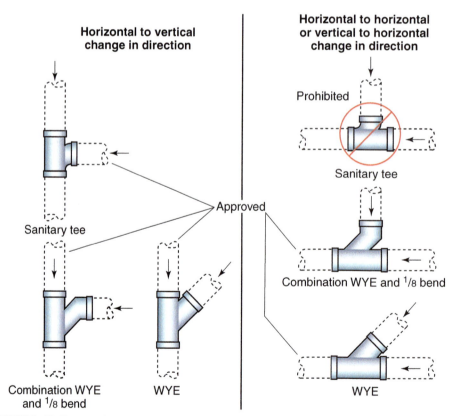

FIGURE 13-4 DWV fittings for change in direction

Cleanouts

The IRC requires cleanouts at locations most susceptible to blockage and situated to accommodate drain cleaning equipment. Cleanouts are required in horizontal drain lines at each change of direction that is greater than 45 degrees. Where more than one change of direction occurs, only one cleanout is required in each 40 feet. Accessible cleanouts are also required within 10 feet of the junction of the building drain and building sewer. A readily removable fixture, such as a water closet or a fixture trap of a sink, may serve as a cleanout for other than the building sewer and building drain junction. Cleanouts must allow cleaning in the direction of the flow and provide working space clearance of at least 18 inches for pipes of 3 inches diameter or more and 12 inches for smaller pipes (Figure 13-5). [Ref. P3005.2]

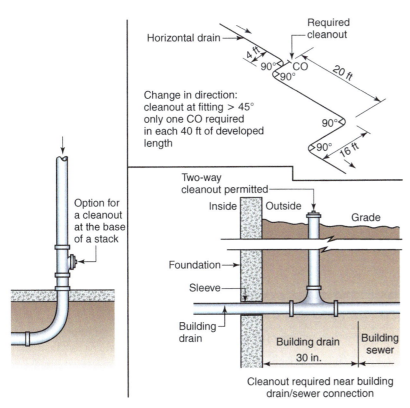

FIGURE 13-5 Cleanout locations

Minimum slope for horizontal drainage piping

The general rule for minimum pipe slope is ¼-inch vertical drop for each 12 inches of horizontal run (¼:12 or 2 percent). A minimum slope of ⅛ inch per foot (⅛:12 or 1 percent) is permitted for pipe sizes 3 inches or larger. [Ref. P3005.3]

Drain pipe sizing

In determining minimum pipe size, the IRC assigns a drainage fixture unit (d.f.u.) value to each plumbing fixture or group of fixtures. Each section of drainage pipe must be sized based on the total d.f.u of the fixtures draining through that section. The code provides separate tables; one for branches and stacks and one for the building drain and its branches. When compared with the tabular values for horizontal branches, values in the building drain table allow a greater d.f.u. capacity for any given size of pipe. Building drain capacity also increases as the slope increases up to a maximum slope of ½ inch per foot. Drain piping serving water closets must be at least 3 inches in diameter (Tables 13-4 through 13-6). **[Ref. P3005.4, Table P3004.1, Table P3005.4.1, Table P3005.4.2]**

EXAMPLE

Using the values in Tables 13-4 through 13-6, determine drainage fixture units and pipe size for each section of sanitary drainage pipe in Figure 13-6.

TABLE 13-4 Drainage fixture unit (d.f.u.) values for various plumbing fixtures

Type of fixture or group of fixtures	d.f.u. value
Bathtub, tub/shower, whirlpool or shower	2
Floor drain	0
Lavatory	1
Full-bath group with bathtub (with 1.6 gal. per flush water closet and with or without shower head and/or whirlpool attachment on the bathtub or shower stall)	5
Half-bath group (1.6 gal. per flush water closet plus lavatory)	4
Kitchen group (dishwasher and sink with or without garbage grinder)	2
Laundry group (clothes washer standpipe and laundry tub)	3
Multiple-bath groups:	
1.5 baths	7
2 baths	8
2.5 baths	9

[Ref. Table P3004.1]

TABLE 13-5 Maximum fixture units allowed to be connected to branches and stacks

Nominal pipe size (in.)	Any horizontal fixture branch	Any one vertical stack or drain
1½	3	4
2	6	10
2½	12	20
3	20	48

[Ref. Table P3005.4.1]

Note: Water closets are not permitted on drain lines smaller than 3 inches.

TABLE 13-6 Maximum fixture units allowed to be connected to building drain, building drain branches or building sewer

Diameter of pipe (in.)	Slope per ft		
	⅛ in.	¼ in.	½ in.
2	–	21	27
3	36	42	50

[Ref. Table P3005.4.2]

Note: Water closets are not permitted on drain lines smaller than 3 inches.

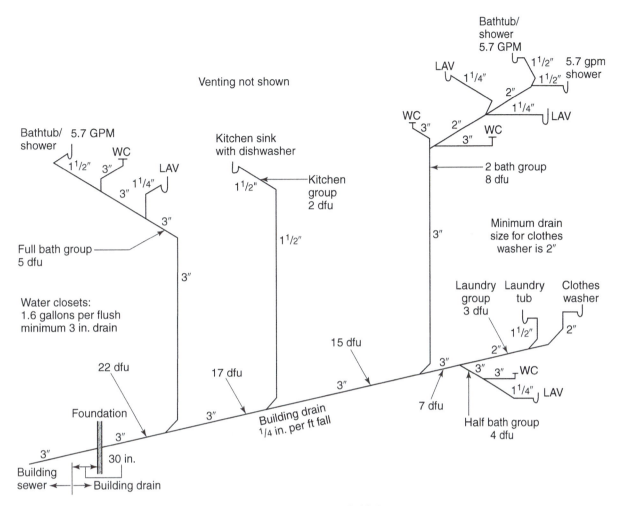

FIGURE 13-6 Drainage pipe sizing based on Tables 13-4 through 13-6

VENTING SYSTEMS

Venting provides air to equalize pressure for proper operation of the drainage system, prevents siphoning of trap seals, and allows the safe escape of sewer gases to the outside atmosphere. To protect the trap seal, the IRC requires venting for every trap and trapped fixture. At least one vent pipe to the outdoors is required and must be sized based upon the size of the building drain.

Vent connections and grades

Vents require adequate slope and alignment to allow moisture and condensation to drain back to the soil and waste pipes. To prevent a stoppage in a dry vent, which would likely go unnoticed, the dry vent connection to a horizontal drain must be above the centerline of the drain. If plumbing is roughed in for future fixtures, a vent must be installed to serve those fixtures. [Ref. P3104]

Fixture vents

For other than self-siphoning fixtures such as water closets, the IRC limits the distance from the trap to the vent and the slope of the fixture drain in order to keep the vent open above the flow line. The total fall of a fixture drain to the vent connection cannot exceed one pipe diameter, and the vent connection is not permitted to be below the trap weir (Table 13-7 and Figure 13-7). **[Ref. P3105]**

TABLE 13-7 Maximum distance of fixture trap from vent

Size of trap (in.)	Slope (in. per ft)	Distance from trap (ft)
1¼	¼	5
1½	¼	6
2	¼	8

[Ref. Table P3105.1]

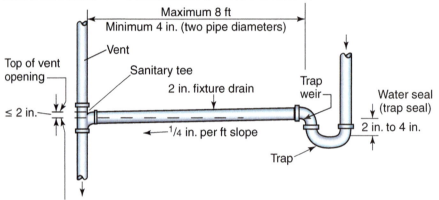

FIGURE 13-7 Fixture vent

Wet venting

Wet venting is based on the principle that pipe serving as a drain for a fixture or fixtures will have adequate amounts of air to concurrently serve as the vent for downstream fixtures. The section of drainage pipe serving as a wet vent must be adequately sized and is limited to a lower d.f.u. load than would otherwise be allowed for drainage piping. Horizontal wet venting is permitted for fixtures of one or two bathroom groups located on the same floor (Table 13-8 and Figure 13-8). **[Ref. P3108]**

TABLE 13-8 Wet vent size

Wet vent pipe size (in.)	Drainage fixture unit (d.f.u.) load
1½	1
2	4
3	12

[Ref. Table P3108.3]

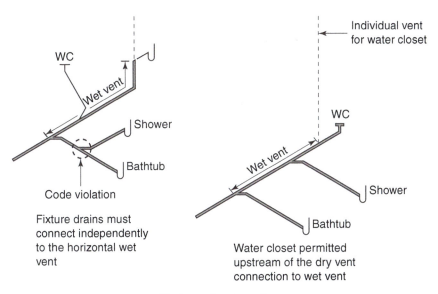

Horizontal wet venting

FIGURE 13-8 Wet venting

Island fixture venting

Island venting is permitted for lavatories and kitchen sinks, including connections for dishwashers and food waste disposers (Figure 13-9). [Ref. P3112]

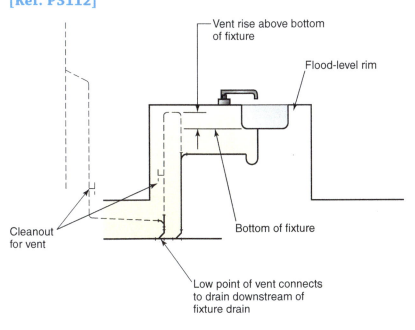

Island fixture vent

FIGURE 13-9 Island venting

Vent pipe sizing

To provide adequate volume of air to the drainage system, the IRC requires the diameter of vent piping to be at least one-half of the required diameter of the drain served, but never less than 1¼ inches. Vents greater than 40 feet long require an increase of one pipe size. [Ref. P3113]

Vent termination

For sloped roofs not used for other purposes, vent pipes must terminate not less than 6 inches above the roof and 6 inches above the anticipated snow accumulation (Figure 13-10). To prevent noxious gas or odors from entering the building, the IRC prescribes minimum clearances between vent terminations and doors, windows or air intake openings (see Chapter 10). [Ref. P3103]

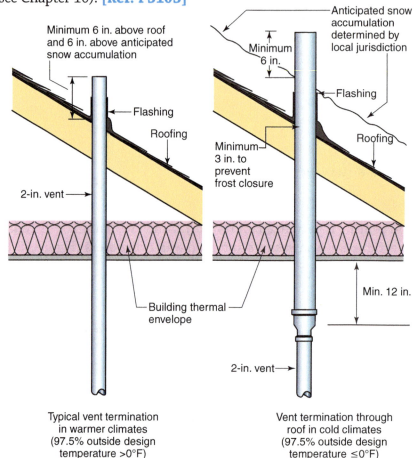

Minimum 6 in. above roof and 6 in. above anticipated snow accumulation

Flashing

Roofing

2-in. vent

Building thermal envelope

Anticipated snow accumulation determined by local jurisdiction

Minimum 6 in.

Flashing

Roofing

Minimum 3 in. to prevent frost closure

Min. 12 in.

2-in. vent

Typical vent termination in warmer climates (97.5% outside design temperature >0°F)

Vent termination through roof in cold climates (97.5% outside design temperature ≤0°F)

FIGURE 13-10 Plumbing vent termination

Air admittance valves

An air admittance valve is a one-way valve designed to allow air into the plumbing drainage system as water drains from the fixture, creating a negative pressure in the piping. The device pulls air from the room in which the fixture is located and, except for the one required through-the-roof vent, is an alternative to venting to the outside air. The valve is designed to close by gravity and seal the terminal under no-flow conditions to prevent introduction of sewer gas into the dwelling. Access is required to the air admittance valve. Use of air admittance valves may be an alternative to island venting of kitchen sinks or lavatories, or for venting other isolated fixtures (Figure 13-11). [Ref. P3114]

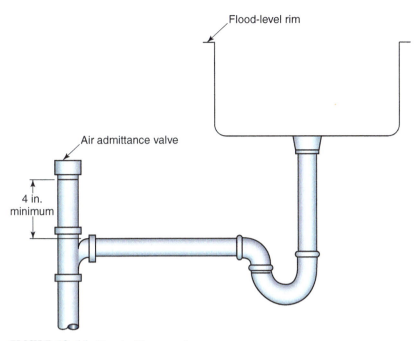

FIGURE 13-11 Air admittance valve

PLUMBING FIXTURES

Fixtures, faucets and fixture fittings must have smooth, impervious surfaces and comply with the applicable referenced standards. The IRC includes requirements for receptors, strainers, valves and features to promote usable and sanitary fixtures and prevent contamination of the potable-water supply.

Laundry standpipes

Standpipe design accommodates the high rate of discharge from clothes washers. Standpipes must extend at least 18 inches and not more than 42 inches above the trap weir (Figure 13-12). **[Ref. P2706.1.2]**

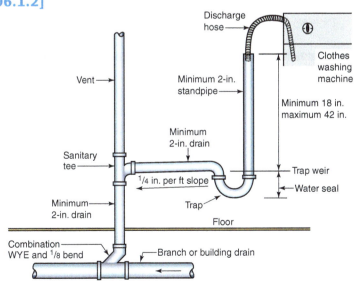

FIGURE 13-12 Laundry standpipe

Dishwashers

The IRC provisions allow a kitchen sink, food waste disposer and dishwasher to discharge through a single 1½-inch trap. The dishwasher drain may discharge through the food waste disposer or it may connect to the sink tailpiece with a wye fitting. The dishwasher waste line must rise to the underside of the counter to minimize the potential for wastewater backflow into the dishwasher (Figure 13-13). **[Ref. P2717]**

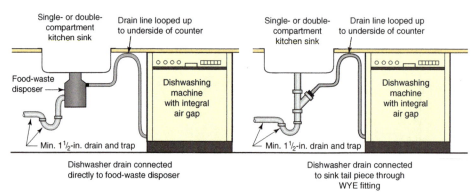

FIGURE 13-13 Dishwasher drain

Protection against scalding

The IRC requires approved temperature control devices on the water outlets of bathing fixtures and bidets to prevent scalding (Table 13-9). **[Ref. P2708.4, P2713.3, P2721.2]**

TABLE 13-9 Temperature control at fixture water supply outlets

Fixture	Max. temperature	Approved device	Standard
Shower or tub/shower combination	120°F	Pressure-balance control valve or Thermostatic-mixing control valve or Combination pressure-balance/thermostatic-mixing control valve	ASSE 1016/ASME A112.1016/CSA B125.16
Bathtub or whirlpool bathtub	120°F	Water-temperature-limiting device	ASSE 1070 or CSA B125.3
Bidet	110°F		

Showers

Minimum shower compartment dimensions are 30 inches by 30 inches. If the shower area is at least 1,300 square inches, the minimum width may be reduced to 25 inches. Hinged shower doors must open outward, with a finished access width of at least 22 inches (see Chapter 10). **[Ref. P2708]**

Whirlpool bathtubs

Adequate access is necessary to service or replace the whirlpool circulation pump. The means and dimensions for access must follow the manufacturer's installation instructions and the code requirements. In all cases, the access opening must be unobstructed and large enough to remove the pump (Figure 13-14). **[Ref. P2720]**

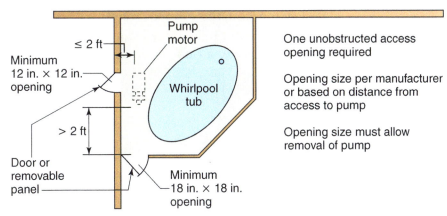

FIGURE 13-14 Two options for access to whirlpool pump

FIXTURE TRAPS

Traps provide a water seal with a depth of 2 inches to 4 inches to prevent sewer gases from entering the building. Floor drains require a trap-primer or deep-seal design to prevent the loss of their water seal by evaporation (Figure 13-15). **[Ref. P3201.2]**

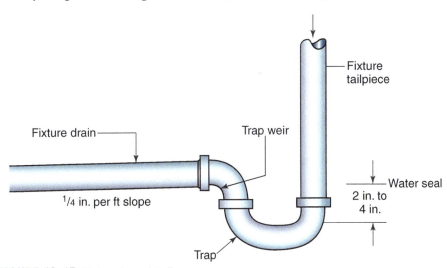

FIGURE 13-15 Fixture trap details

WATER HEATERS

The IRC prescribes location, installation, and safety requirements for water heaters in the plumbing, fuel-gas, and electrical chapters of the code. The plumbing section covers connections to the water supply, drain pan installations, and temperature relief valve provisions. As in the fuel-gas provisions, water heaters installed in garages must have ignition sources elevated at least 18 inches above the garage floor unless the water heater is listed as flammable-vapor-ignition resistant. For buildings located in seismic design categories (SDCs) D_0, D_1 and D_2, and townhouses located in SDC C, water heaters require anchorage to the walls of the structure to resist horizontal earthquake forces. Where water leaking from the storage tank will cause damage, the IRC requires the water heater to be installed in a pan that drains to an approved location, typically a floor drain or other indirect receptor (Figure 13-16). **[Ref. P2801, P2804]**

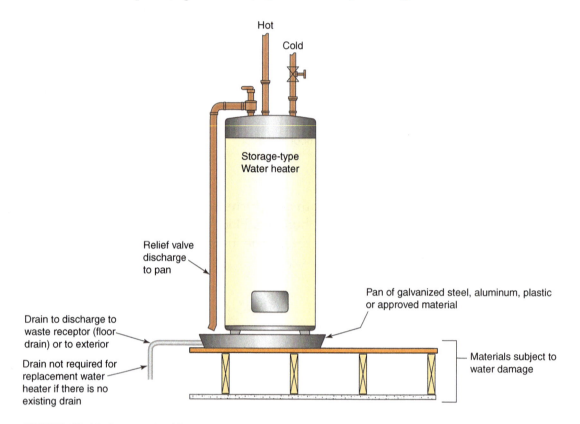

FIGURE 13-16 Pan required for storage-type water heaters where leaks will cause damage

CHAPTER
14

Electrical

The electrical chapters of the *International Residential Code* (IRC) are based on the 2017 *National Electrical Code* (NEC) (NFPA 70-2017), published by the National Fire Protection Association (NFPA). The IRC provisions cover electrical installations normally occurring in the construction of one- and two-family dwellings and townhouses. This does not prevent the use of other wiring methods and materials that conform to the NEC requirements. Discussion in this chapter will focus on commonly encountered electrical installations for services, branch circuits, devices and fixtures in IRC-regulated buildings.

ELECTRICAL SERVICES

The scope of the IRC covers services for 120/240-volt, single-phase systems not more than 400 amperes. The serving utility company delivers electrical power to the service entrance conductors, which in turn conduct the energy to the main service disconnect. The service distributes electricity to the premises wiring system. Only one service is permitted for one- and two-family dwellings. **[Ref. E3601.2]**

Equipment location

The electrical service equipment is typically a cabinet mounted to an interior wall and containing a panelboard with a main disconnect that shuts off all power to the building, the circuit breakers serving the branch circuits and related equipment. The service disconnect must be readily accessible and, when installed inside, must be near to where the service conductors enter the building. Working space in front of equipment, including service panels or subpanels, must be at least 30 inches wide by 36 inches deep with headroom of at least 6 feet 6 inches, and have a light source nearby. The spaces above and below the panel are dedicated to the electrical installation. This means, for example, that pipes or ducts cannot be located directly above the panel. The IRC also does not permit the installation of electrical panels, service disconnects, or circuit breakers in clothes closets or bathrooms (Figure 14-1). **[Ref. E3405, E3601.6]**

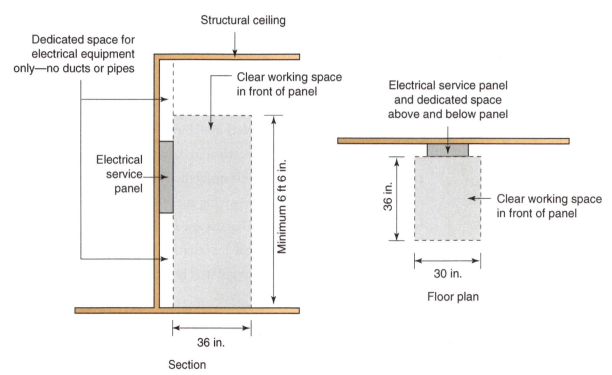

FIGURE 14-1 Electrical panel location and clearance

Electrical service size and rating

The minimum service rating for single-family dwellings is 100 amperes. The code permits a minimum service rating of 60 amperes for other installations. The ampacity of ungrounded service conductors and the rating of the service disconnect may not be less than the load served. The minimum service load for a single-family dwelling is calculated according to Table 14-1. [Ref. E3602]

TABLE 14-1 Minimum service load calculation for a single-family dwelling

Loads and procedure				Volt-amperes
General lighting and general use receptacle outlets	3 volt-amperes	X	Floor area (sq. ft.)	_____ VA
				Plus
All 20-ampere-rated small appliance and laundry circuits	1500 volt-amperes	X	Number of circuits	_____ VA
				Plus
Appliances	The nameplate volt-ampere rating of all permanent or dedicated appliances:			
			Ranges	_____ VA
			Ovens	_____ VA
			Cooking units	_____ VA
			Clothes dryers	_____ VA
			Water heaters	_____ VA
			Subtotal	_____ VA
				Plus
Apply the following demand factors to the above subtotal:	100%	X	First 10,000 volt-amperes	_____ VA
				Plus
	40%	X	Portion in excess of 10,000 volt-amperes	_____ VA
				Plus
Air conditioning	Total of the nameplate rating(s) of the air-conditioning and cooling equipment			_____ VA
			Total volt-amperes	_____ **VA**
Total load in amperes	**Volt-ampere sum**	×	**240 volts =**	_____ **AMPS**

Service conductor size

In the more precise language of the code, hot current-carrying wires are referred to as *ungrounded conductors,* and the neutral wires are referred to as *grounded conductors.* Table 14-2 identifies the minimum size of the ungrounded service conductor to provide the required ampacity based on the wire material and type of insulation. The grounded (neutral) conductor ampacity must be at least the maximum unbalance of the load and its size not less than the required minimum grounding electrode conductor size. Though in practice a rule of thumb has held that the neutral may be as much as two sizes smaller than the ungrounded service conductors, the code requires that the size be calculated. The grounding electrode system is discussed later in this chapter, but the size of the grounding electrode conductor is based on the size of the service entrance conductors, as shown in Table 14-2. [Ref. E3603, Table E3603.4]

TABLE 14-2 Service conductor and grounding electrode conductor size

Conductor types and sizes*		Allowable ampacity, max. load (amps)	Min. grounding electrode conductor size	
Copper (awg)	Aluminum and copper-clad aluminum (awg)		Copper (awg)	Aluminum (awg)
4	2	100	8	6
3	1	110	8	6
2	1/0	125	8	6
1	2/0	150	6	4
1/0	3/0	175	6	4
2/0	4/0 or two sets of 1/0	200	4	2
3/0	250 kcmil or two sets of 2/0	225	4	2
4/0 or two sets of 1/0	300 kcmil or two sets of 3/0	250	2	1/0
250 kcmil or two sets of 2/0	350 kcmil or two sets of 4/0	300	2	1/0
350 kcmil or two sets of 3/0	500 kcmil or two sets of 250 kcmil	350	2	1/0
400 kcmil or two sets of 4/0	600 kcmil or two sets of 300 kcmil	400	1/0	3/0

*THHN, THHW, THW, THWN, USE, RHH, RHW, XHHW, RHW-2, THW-2, THWN-2, XHHW-2, SE, USE-2.

Note: Service conductors in parallel sets of 1/0 and larger are permitted in either a single raceway or in separate raceways. Grounding electrode conductors of size 8 AWG require protection with conduit. Grounding electrode conductors of size 6 AWG require protection with conduit or must closely follow a structural surface for physical protection.

GROUNDING

The grounding system provides a fault current path to earth. Grounding conductors from equipment, enclosures, and devices are bonded to the grounded (neutral) conductor and grounding electrode conductor at the service equipment. The electrical system is connected to the earth through the grounding electrode conductor(s) connected to one or more approved grounding electrodes in contact with the earth. [Ref. E3607]

Grounding electrode system

Commonly used grounding electrodes include underground metal water pipe, concrete-encased reinforcing bar and approved ground rods. The electrodes, when present, are bonded together to form the grounding electrode system. An electrode of underground metal water pipe must have at least 10 feet in contact with the ground, with connection to the grounding electrode conductor within 5 feet of entrance into the building. Interior metal water piping more than

5 feet from the entrance into the building is not considered part of the grounding electrode system and cannot be used as a conductor to connect electrodes. Where underground metal water pipe is available and used as an electrode, the code requires at least one additional electrode.

Concrete-encased grounding electrodes (often called an *Ufer ground*) have proven very effective and do not require supplemental electrodes. The effectiveness of this method relies on the conductive characteristics of the concrete, the increased surface area in contact with the soil, and the ability of concrete to absorb and retain moisture. This type of electrode consists of 20 feet of ½-inch reinforcing bar or 20 feet of 4 AWG bare copper wire encased in a concrete footing or foundation in contact with the ground. At least 2 inches of concrete cover is required. If such reinforcing is present in a concrete footing, the code requires it to be used as part of the grounding electrode system, though no more than one concrete encased electrode is required.

One or more rod, pipe, ring or plate grounding electrodes, sometimes referred to as "made" electrodes, are required under two circumstances: where there are no other electrodes available or where underground water pipe is the only other grounding electrode present. This requirement is typically satisfied with a listed copper-clad ground rod 8 feet long driven into the soil. Unless testing after installation determines that a single ground rod has a resistance of 25 ohms or less, the code requires two such electrodes spaced not less than 6 feet apart.

Grounding electrode conductors are sized based on the size of the ungrounded service conductors, as shown in Table 14-2. The code generally does not permit splicing of grounding electrode conductors run to one or more grounding electrodes. Connections to electrodes require approved fittings to ensure an effective grounding path. Grounding electrode conductors require protection in conduit or by other approved means if exposed to potential physical damage (Figure 14-2). [Ref. E3608, E3610, E3611]

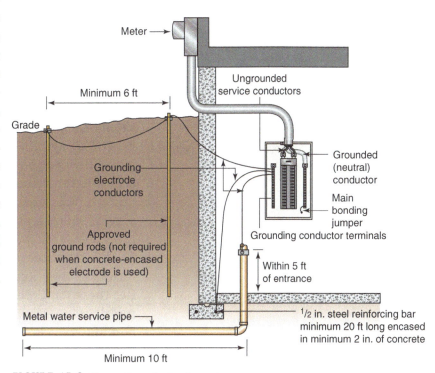

FIGURE 14-2 Grounding electrode system

Bonding

The code provides for the bonding of metal parts associated with the electrical system to provide an effective path for fault current. A main bonding jumper, such as the green machine screw or green insulated conductor supplied with the service enclosure, is required to connect the grounded (neutral) conductors to the service equipment enclosure and the equipment grounding conductors. As a general rule, the main service disconnect enclosure is the only location where the code permits connection of the grounding system to the grounded (neutral) conductors. They are isolated from each other elsewhere to prevent creating a parallel fault path.

Bonding of the metal water piping system to the service equipment enclosure with an appropriately sized bonding conductor is required in order to provide a fault path should the water pipe become energized through accidental contact with hot wires. The code does not permit the interior water piping system to be used as a ground, a grounding electrode, or a conductor for the grounding electrode system. The points of attachment of the bonding jumper must be accessible (Figure 14-3). [Ref. E3609]

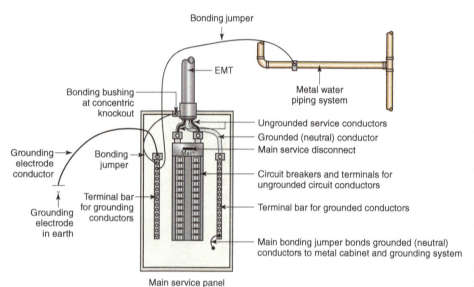

FIGURE 14-3 Bonding at the service panel

Bonding for other systems

For bonding of other systems—typically telephone, satellite, and cable television systems—to the building grounding system, the code requires installation of an *intersystem bonding termination*, a device that provides a means for connecting grounding and bonding conductors of communications systems near the building service equipment. The bonding termination must be accessible and must have provisions for connection of at least three bonding conductors. A set of listed terminals mounted directly to the meter cabinet satisfies these requirements and may be the most common installation. Other approved methods for intersystem bonding include the installation of a bonding bar near the service enclosure, meter cabinet, or service raceway (Figure 14-4). [Ref. E3609.3]

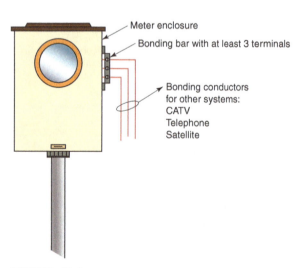

FIGURE 14-4 Intersystem bonding termination

BRANCH CIRCUITS

The IRC specifies the minimum required branch circuits, circuit ratings, conductor size, and overcurrent protection. The total number of branch circuits is determined from the total calculated load.

Branch circuit ratings and conductor size

Branch circuits must be rated according to the maximum allowable ampere rating or setting of the overcurrent protection device, typically a circuit breaker. Branch circuit requirements are summarized in Table 14-3. [Ref. E3702, Table E3702.14]

TABLE 14-3 Branch circuit requirements—summary

Conductors	Circuit rating		
	15 amp	20 amp	30 amp
Min. size (AWG) circuit conductors (copper)	14	12	10
Overcurrent-protection device: max. amp rating	15	20	30
Duplex or multiple outlet receptacle rating (amps)	15 max.	15 or 20	30
Single receptacle outlet minimum rating (amps)	15	20	30
Max. load (amps)	15	20	30

[Ref. Table E3702.14 and Section E4002.1]

Conductor sizing

Table 14-3 contains the basic conductor sizing criteria applicable to the majority of wiring in a dwelling. For example, Type NM (nonmetallic) cable with 12 AWG copper conductors is typically satisfactory for all 20-amp branch circuits. The code also provides ampacity tables for all wire sizes based on the material and insulation type of the conductor. When sizing wires using the ampacity tables, typically for circuits exceeding 30 amps, a number of variables must be considered, including the required temperature rating of the conductor insulation, derating for bundled conductors, and the temperature rating of the terminal. [Ref. E3705]

Overcurrent protection required

An overcurrent device, either a circuit breaker or fuse, is required to protect all ungrounded branch circuit and feeder conductors. Overcurrent protective device ratings or settings cannot exceed the allowable ampacity of the conductor. When the current passing through the conductor exceeds the ampacity rating due to overload, short circuit or ground fault, the overcurrent device opens the circuit before heat buildup can cause damage or a fire (Table 14-4). [Ref. E3705.5, Table E3705.5.3]

TABLE 14-4 Overcurrent protection rating

Copper		Aluminum or copper-clad aluminum	
Size (awg)	Max. overcurrent protection device rating (amps)	Size (awg)	Max. overcurrent protection device rating (amps)
14	15	12	15
12	20	10	25
10	30	8	30

[Ref. Table E3705.5.3]

Note: The maximum overcurrent protection device rating shall not exceed the conductor's allowable ampacity after applying any correction or adjustment factors.

Location of overcurrent devices

Overcurrent devices are installed at the point where the branch circuit conductors receive their supply, typically at the service panel. Such devices must be readily accessible and not subject to damage. The code prohibits overcurrent devices and panelboards in clothes closets or bathrooms, or located above a step. [Ref. E3705.7, E3405.4]

Required branch circuits

The minimum required branch circuits are summarized in Table 14-5. [Ref. E3703]

TABLE 14-5 Required branch circuits

Area or appliance served	Circuit(s)	Circuit may also serve	Alternate
Central heating equipment	Individual branch circuit sized for equipment	–	–
Kitchen countertop receptacles	Min. two 20-amp small appliance circuits	Other receptacle outlets in the kitchen, pantry, breakfast and dining areas	Refrigerator may be served by an individual 15-amp circuit
All laundry area receptacles	Separate 20-amp circuit	–	–
Bathroom receptacle outlet(s)	Min. one 20-amp circuit	Receptacle outlets in multiple bathrooms	A circuit dedicated to a single bathroom may supply lights, fan and other equipment in that bathroom
Garage receptacle outlets	Min. one 20-amp circuit	Readily accessible outdoor receptacle outlet	–

WIRE AND TERMINAL IDENTIFICATION

In general, electrical devices common to residential construction, such as receptacles, that connect to both the ungrounded (hot) and grounded (neutral) conductors require properly identified terminals. The identification requirement does not apply to panelboards or devices rated over 30 amps. [Ref. E3407]

Grounded (neutral) conductors and terminals

Grounded conductors not more than 6 AWG are identified by white or gray insulation or have three white stripes on other than green insulated wire. Grounded conductors greater than 6 AWG may have the same markings, but the code also permits other colors of insulation, typically black, provided the electrician applies white or gray tape or marking around the conductor at the terminal ends at the time of installation. The grounded conductor terminal on a receptacle must be substantially white in color or identified by the word "white" or letter "W." In practice, grounded conductor terminals are typically silver-colored metal (designating white), and ungrounded terminals are a darker, brass-colored metal. **[Ref. E3407.1, E3407.4]**

Grounding conductors

Grounding conductors have green insulation or yellow stripes on green insulation, or are bare wires. For grounding conductors greater than 6 AWG, the code permits insulation of another color, typically black, provided the electrician applies green tape or marking around the conductor at the terminal ends and access points at the time of installation. As an alternative, the insulation may be stripped to bare wire for the entire exposed length. **[Ref. E3407.2]**

Ungrounded (hot) conductors

Ungrounded conductors generally must have insulation of any color other than white, gray, or green. **[Ref. E3407.3]**

WIRING METHODS

The IRC recognizes a number of approved wiring methods, including armored cable, metal-clad cable and individual conductors installed in various types of metallic and nonmetallic conduit. Discussion in this chapter will focus on above-ground wiring methods using Type NM non-metallic sheathed cable common to construction of one- and two-family dwellings and townhouses. In all cases, the cable must be approved for the location. For example, Type NM cable is not permitted underground, may not be used in wet or damp locations, and is not to be embedded in concrete. **[Ref. E3801.4, Table E3801.4]**

Type NM cable

The installation requirements for Type NM cable are mainly concerned with protection from physical damage, including penetration from fasteners. The required setbacks from the edges of framing members and minimum cable support requirements are shown in Table 14-6. In unfinished basements, Type NM cable smaller than 8 AWG is prohibited from being attached directly to the bottom of joists (Figures 14-5 through 14-7). **[Ref. E3802, Table E3802.1]**

TABLE 14-6 Summary of installation requirements for Type NM* cable

Installation	Physical protection	
	Minimum setback from edge of framing (in.)	**Physical protection if minimum distance is not met**
Cable run parallel with the framing member or furring strip	1¼	0.0625-inch steel plate or sleeve
Cable in bored holes or notches in framing members	1¼	0.0625-inch steel plate or sleeve
Cable installed in grooves and covered	1¼ free space	0.0625-inch steel plate or sleeve
Support of cable		
Maximum allowable on center support spacing	4.5 feet	
Maximum support distance from metal box with cable clamp	12 inches	
Maximum support distance from plastic box without cable clamp	8 inches	
Flat cables shall not be stapled on edge	—	
Conductor length		
Minimum free conductor length at each box	6 inches with sheath removed and 3 inches outside of box	

*NM = nonmetallic sheathed cable.

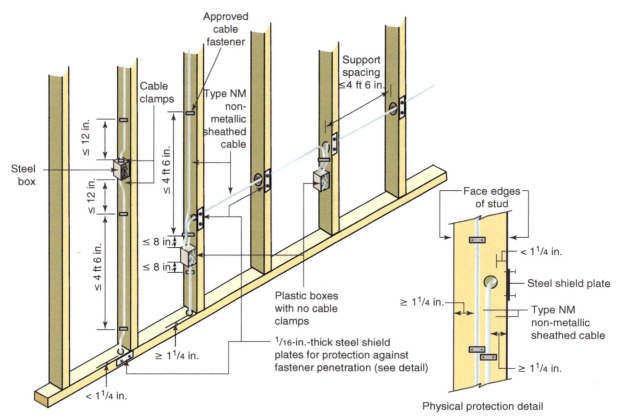

FIGURE 14-5 Type NM cable installation

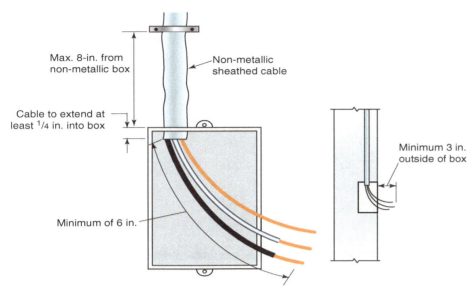

Max. 8-in. from
non-metallic box

Non-metallic
sheathed cable

Cable to extend at
least ¼ in. into box

Minimum 3 in.
outside of box

Minimum of 6 in.

FIGURE 14-6 Free conductor length at boxes

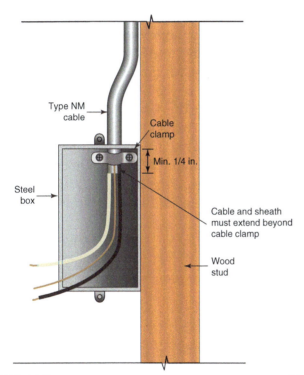

Type NM
cable

Cable
clamp

Min. 1/4 in.

Steel
box

Cable and sheath
must extend beyond
cable clamp

Wood
stud

FIGURE 14-7 Non-metallic sheathed cable entering a metal box

Underground installation requirements

Cable installed underground must be listed and labeled for the location. In addition, direct burial conductors and cables emerging from the ground must be protected from 18 inches below grade to 8 feet above grade using the prescribed methods of the code. Splices underground are permitted only with specific approved methods and listed materials (Table 14-7). **[Ref. E3803]**

TABLE 14-7 Minimum cover requirements, burial (in.)

Location of wiring method or circuit	Type of wiring method or circuit		
	Direct burial cables or conductors	Rigid metal conduit or intermediate metal conduit	Non-metallic raceways listed for direct burial without concrete encasement or other approved raceways
All locations not specified below	24	6	18
Under a building	0 (in raceway only)	0	0
Under streets, highways, roads, alleys, driveways and parking lots	24	24	24
One- and two-family dwelling driveways and outdoor parking areas, and used only for dwelling-related purposes	18	18	18

[Ref. Table E3803.1]

Code Essentials

Requirements for boxes supporting ceiling paddle fans

- Marked by manufacturer as suitable
- Marked for maximum weight when supporting 36- to 70-pound fan
- Not permitted for fans over 70 pounds

Requirements for boxes supporting luminaires (lighting fixtures)

- Designed for the purpose
- Capable of supporting 50 pounds
- Permitted for luminaires weighing 50 pounds or less
- Not permitted for luminaires weighing over 50 pounds unless the box is listed and marked for the weight ●

Boxes

A box is required at each splice, junction, outlet, switch and pull point. All metal boxes require grounding by approved means, typically a short bonding jumper between a bonding screw on the metal box and the equipment grounding conductor. Each box is limited to a maximum box fill capacity based on the size of the box, the number and size of conductors entering the box, and the devices and fittings installed. Unused cable openings, typically created by the removal of a knockout plate, must be closed by approved methods to provide protection equivalent to the wall of the box. The code places limitations on boxes used for supporting ceiling-suspended paddle fans and luminaires (lighting fixtures) (Figure 14-8). **[Ref. E3905, E3906.4, E4101.6]**

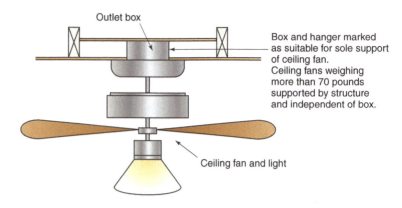

Outlet box

Box and hanger marked as suitable for sole support of ceiling fan. Ceiling fans weighing more than 70 pounds supported by structure and independent of box.

Ceiling fan and light

FIGURE 14-8 Box supporting ceiling fan

Grounding

Metal enclosures, devices and equipment require grounding in accordance with the code. Equipment grounding conductors provide a path to ground through the equipment grounding and grounding electrode system. Type NM cable requires an insulated or bare grounding conductor used for no other purpose. **[Ref. E3908]**

POWER DISTRIBUTION

The IRC establishes location requirements for receptacle and lighting outlets.

Receptacle outlet locations

In addition to providing convenience for the occupants, proper placement of receptacle outlets to serve fixtures, tools and appliances reduces electrical hazards in the home, such as may occur with the use of extension cords (Table 14-8 and Figures 14-9 through 14-16). [Ref. E3901]

TABLE 14-8 Receptacle outlet locations

Description	Maximum spacing (ft)	Minimum number of receptacle outlets	Location	Note
Habitable rooms	12		within 6 ft of a door and within 6 ft of any point along a wall	for wall space ≥ 24 in. wide; wall receptacles ≤ 5.5 ft above the floor; floor receptacles ≤ 18 in. from wall
Kitchen wall counters	4		within 2 ft of any point along the wall line and ≤ 20 in. above counter	for wall counter ≥ 12 in. wide
Kitchen island or peninsular counter		1	≤ 20 in. above counter	for counter ≥ 12 in. × 24 in.
Hallway		1		for hallways ≥ 10 ft
Foyer		1	each wall ≥ 3 ft	for foyers > 60 sq. ft.
Bathroom		1	≤ 36 in. from lavatory	measured from the outside edge of each lavatory basin, located above or < 12 in. below top of basin
Outdoors		1 in front & 1 in back	≤ 6 ft 6 in. above grade	where there is access to grade
Deck, balcony or porch		1	≤ 6 ft 6 in. above floor	applies to all decks, balconies, or porches accessible from inside dwelling
Laundry areas		1	≤ 6 ft from appliance	–
Basement		1	–	for basement with finished habitable space, one outlet for each separate unfinished area
Garage		1	each vehicle bay	min. one outlet per vehicle space
HVAC equipment		1	≤ 25 ft from equipment	–

HVAC = Heating, ventilation, air conditioning

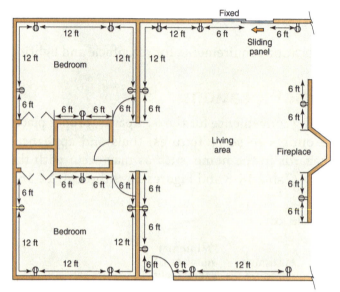

Dimensions shown are maximum spacing of general purpose receptacles

FIGURE 14-9 Habitable room receptacle outlet locations

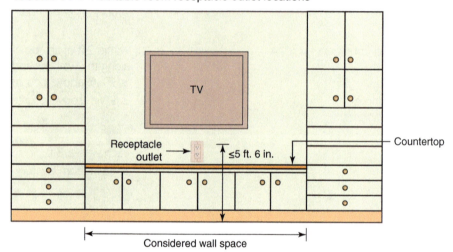

Elevation—Family room

FIGURE 14-10 Cabinets with countertops or work surfaces are counted as wall space

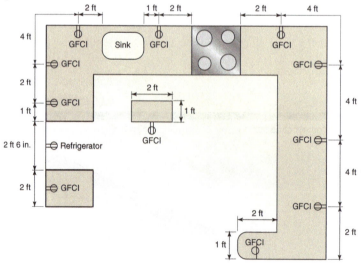

FIGURE 14-11 Kitchen counter receptacle outlets

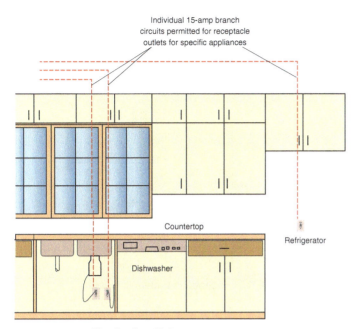

Individual 15-amp branch
circuits permitted for receptacle
outlets for specific appliances

Countertop

Refrigerator

Dishwasher

Elevation view—Kitchen

FIGURE 14-12 Receptacle outlet for a specific appliance can be on an individual 15-amp branch circuit

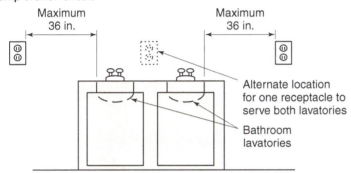

Maximum
36 in.

Maximum
36 in.

Alternate location
for one receptacle to
serve both lavatories

Bathroom
lavatories

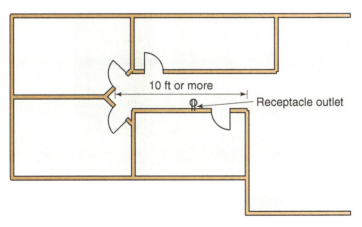

10 ft or more

Receptacle outlet

FIGURE 14-13 Bathroom and hallway receptacle outlets

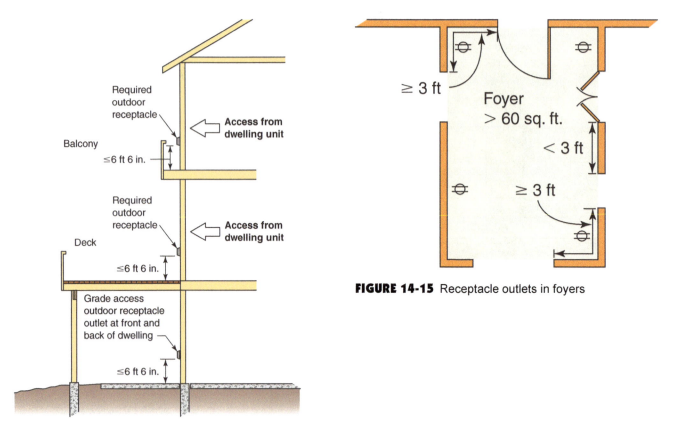

FIGURE 14-14 Outdoor receptacle outlets

FIGURE 14-15 Receptacle outlets in foyers

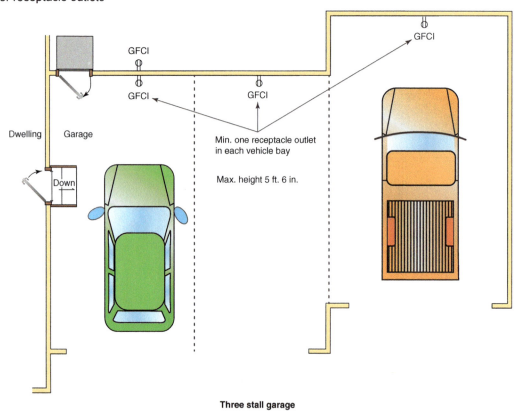

Three stall garage

FIGURE 14-16 Receptacle outlet required in each vehicle space in garage

Lighting outlets

Wall switch-controlled lighting outlets are required in all habitable rooms, bathrooms, hallways, storage areas and garages, at each stairway and outside each exterior door (Figure 14-17). [Ref. E3903]

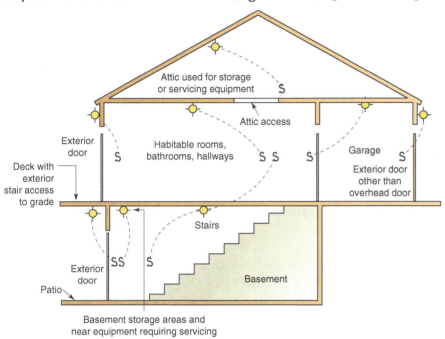

FIGURE 14-17 Wall switch–controlled lighting outlets

Ground-fault circuit interrupter (GFCI) protection

GFCI devices protect people from shock hazards by de-energizing a circuit or receptacle when a fault current to ground is detected. Receptacle outlets of 125 volts, 15 and 20 amps, require GFCI protection at the locations shown in Figure 14-18. [Ref. E3902]

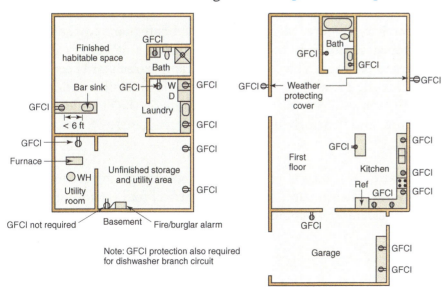

FIGURE 14-18 GFCI-protected receptacle outlets

Arc-fault circuit interrupter (AFCI) protection

AFCI devices detect unwanted arcing in the wiring of the branch circuit and open the circuit before excessive heat buildup can cause a fire. The AFCI device typically is installed in the service panel or subpanel in order to protect the entire branch circuit. AFCI protection is required for branch circuits serving 120-volt, single-phase, 15- and 20-ampere receptacle, lighting and smoke alarm outlets in most living areas of a dwelling unit, including hallways and closets. Bathroom, unfinished basement, garage and outdoor outlets require GFCI protection but do not require AFCI protection. **[Ref. E3902.15 through E3902.17]**

RECEPTACLES AND LUMINAIRES

The IRC regulates the location and installation of receptacles and luminaires (lighting fixtures) to protect against fire and shock hazards.

Receptacles

FIGURE 14-19 Weatherproof cover

A single receptacle installed on an individual branch circuit must have an ampere rating not less than that of the branch circuit. The ampere rating of receptacles connected to a branch circuit supplying two or more receptacles is determined from Table 14-3. When installed in a wet location, 15- and 20-amp receptacles require enclosures that are weatherproof when a cord is plugged in (Figure 14-19). The code prohibits receptacles within or directly over a bathtub or shower space. **[Ref. E4002]**

Tamper-resistant receptacles

Tamper-resistant receptacles are designed to prevent the insertion of any small object, such as a paper clip, into one side of the receptacle. Both blades of an attachment plug must be inserted simultaneously to open the protective shield and allow connection to electricity (Figure 14-20). This added safeguard in the electrical provisions intends to reduce the number of electrical shock injuries to children. The code requires tamper-resistant receptacles for 125-volt, 15- and 20-ampere receptacles installed in locations accessible to children within dwelling units on the outside of dwelling units, and in attached and detached garages. **[Ref. E4002.14]**

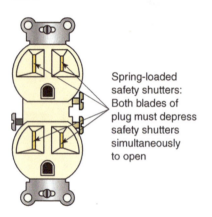

Spring-loaded safety shutters: Both blades of plug must depress safety shutters simultaneously to open

FIGURE 14-20 Tamper resistant receptacles

Luminaires

"Luminaire" is the code term for a complete "lighting fixture." The code regulates luminaires and other electrical fixtures for type, location and clearance to combustibles.

Recessed luminaire installation and clearance

Type IC (insulation contact) recessed luminaires are listed for installation in contact with insulation and combustible materials (see Chapter 15). Other types of recessed luminaires must maintain minimum distances to combustibles and thermal insulation. **[Ref. E4004.8, E4004.9]**

Bathtub and shower areas

Specific requirements apply for fixture installations in a zone measured 3 feet horizontally and 8 feet vertically from the top of a bathtub rim or shower stall threshold, including the area directly above the tub or shower. For example, cord connected, suspended, pendant or track light luminaires and paddle fans are not permitted in this zone. Permitted luminaires located above a bathtub must be marked as suitable for damp locations. Where they are subject to shower spray, they must be marked for wet locations (Figure 14-21). **[Ref. E4003.11]**

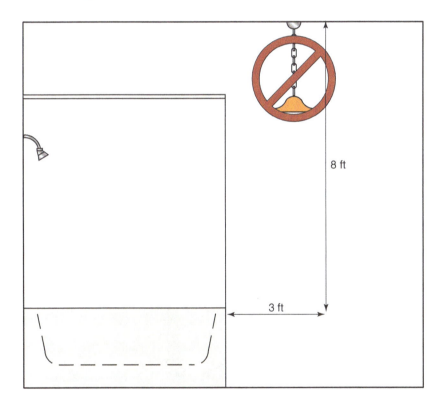

8 ft

3 ft

FIGURE 14-21 Luminaires in bathtub and shower areas

Luminaires in clothes closets

To reduce fire hazard, the code restricts the type of luminaires in clothes closets and sets minimum clearances to the storage areas. Specified clearances are measured from the fixture to the nearest point of the defined storage space (Table 14-9 and Figure 14-22). **[Ref. E4003.12]**

TABLE 14-9 Luminaires in clothes closets

Permitted luminaire	Min. clearance to storage area (in.)	Prohibited luminaires
Incandescent		• incandescent luminaires with open or partially enclosed lamps • pendant luminaires • lamp holders
Surface-mounted incandescent with completely enclosed lamps	12	
Recessed incandescent with completely enclosed lamps	6	
Fluorescent		
Surface-mounted fluorescent	6	
Recessed fluorescent	6	
Surface-mounted fluorescent identified for use in storage space	0*	
LED		
Surface-mounted LED with completely enclosed light source	12	
Recessed LED with completely enclosed light source	6	
Surface-mounted LED identified for use in storage space	0*	

* Luminaire permitted within defined storage space

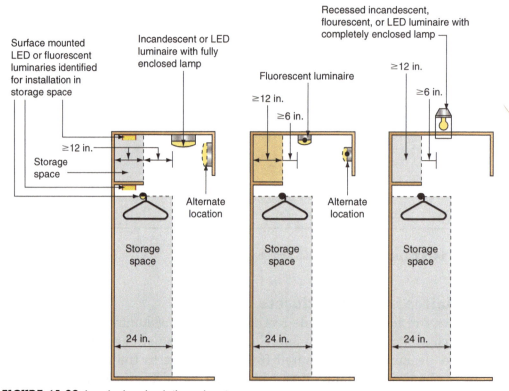

FIGURE 14-22 Luminaires in clothes closets

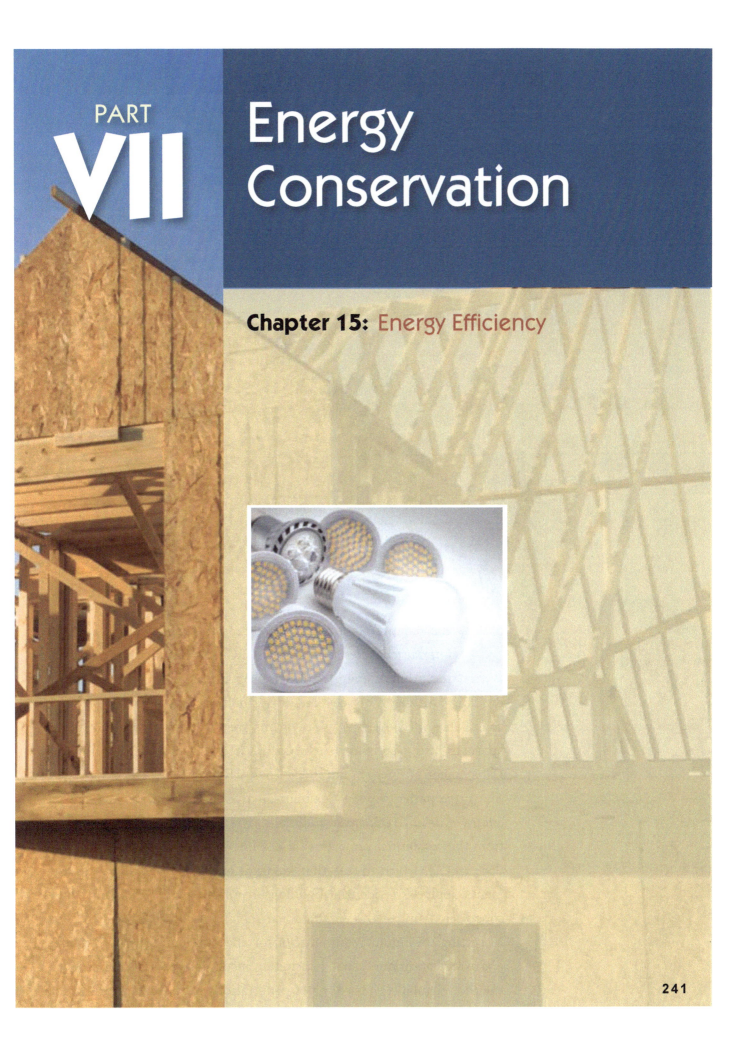

CHAPTER 15

Energy Efficiency

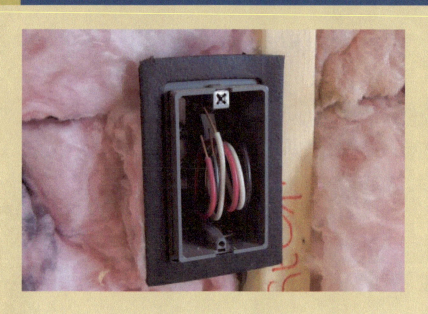

This chapter addresses the prescriptive methods of the *International Residential Code* (IRC) for effective use and conservation of energy through proper design and construction of dwellings. Although not covered in this publication, the other paths for compliance with the energy provisions are 1) a simulated performance approach using software such as REScheck™ available from the U.S Department of Energy and 2) the Energy Rating Index (ERI) approach. The three paths provide flexibility in the design and construction of energy-efficient residential buildings. The IRC provisions are adapted from the residential provisions of the *International Energy Conservation Code* (IECC), which means that identical energy conservation requirements apply to residential buildings regulated by the IRC and those constructed under the *International Building Code* (IBC). In addition to setting minimum requirements for the building thermal envelope enclosing conditioned space, including insulation, windows and doors, the code regulates the sealing of penetrations to reduce air infiltration and the insulation and sealing of ductwork for heating, ventilation and air conditioning (HVAC). Mechanical system controls, insulation of piping systems, and energy-efficient lighting are also covered. Climate zones assigned to geographic location are the basis for specific thermal envelope requirements. The code lists the climate zone designation for each county corresponding to the climate zone map (Figure 15-1).

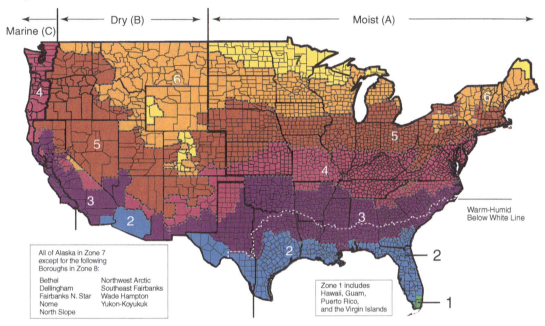

FIGURE 15-1 Climate zone map

BUILDING INSULATION

The energy conservation provisions prescribe methods for identifying components of the thermal envelope to verify compliance with the code. The prescriptive tabular values for minimum insulation R-values for each building component are based on the climate zone.

Insulation identification and verification

Each piece of insulation 12 inches or more in width requires a manufacturer's mark visible after installation that identifies its R-value. As an alternative, the code requires the installer to provide certification stating the type, manufacturer and R-value of the insulation. In addition, for fiberglass or cellulose blown-in or sprayed insulation, the certification must include

- Initial installed thickness
- Settled thickness
- Settled R-value
- Installed density
- Coverage area
- Number of bags installed

You Should Know

R-value

R-value is used to rate the relative thermal resistance to heat flow through insulation. A higher R-value indicates greater resistance and more effective insulation. The type of insulating material, its density, and the installed thickness determine the R-value of thermal insulation. The insulation R-value does not in itself indicate the overall wall, floor or ceiling R-value. The effectiveness of these insulated assemblies depends in part on the method of installation. For example, insulation installed between studs does not improve the resistance to heat flow through those studs. In this case, the overall average wall R-value will be less than the cavity insulation R-value. ●

The insulation installer must sign, date and post the insulation certificate in a conspicuous location. This certification is in addition to the mandatory permanent certificate discussed later in this chapter.

When fiberglass or cellulose insulation is blown or sprayed in the joist or truss spaces of the attic, the IRC requires fixed markers to indicate the installed thickness of the insulation. At least one marker must be installed for every 300 square feet. Minimum 1-inch-high numbers must be visible from the attic access for inspection purposes (Figure 15-2). **[Ref. N1101.10]**

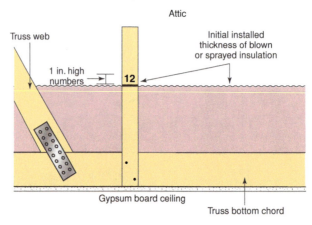

FIGURE 15-2 Attic insulation depth markers

Insulation requirements

Minimum R-values for each insulated component of the thermal envelope are shown in Table 15-1. A number of exceptions to the tabular values exist, as indicated in Table 15-2. For example, the code recognizes the increased efficiency of an energy-type or raised-heel roof truss installed in cold climates to provide the full depth of the insulation above the wall plates and reduces the required ceiling insulation R-value accordingly (Figure 15-3). The code also recognizes the practical difficulties in achieving high R-values when the rafter or joist space is too small to accommodate the required thickness of insulation (Figure 15-4). The values in Tables 15-1 and 15-2 apply to conventional wood frame construction. Due to steel's high thermal conductivity, the IRC requires higher insulation R-values and, in most cases, continuous insulation sheathing to provide a thermal break for steel-frame floor, wall and ceiling construction. **[Ref. N1102.1, N1102.2]**

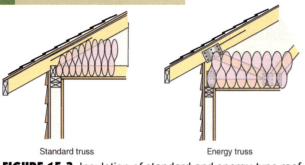

FIGURE 15-3 Insulation of standard and energy-type roof trusses

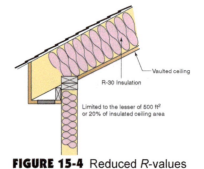

FIGURE 15-4 Reduced R-values for vaulted ceilings

TABLE 15-1 Insulation minimum *R*-value requirements by component

Climate zone	Ceiling *R*-value	Wood frame wall *R*-value	Mass wall *R*-value		Floor *R*-value	Basement and crawl space wall *R*-value		Unheated slab	
			≤50% on interior	>50% on interior		Continuous	Cavity	*R*-value	Depth (ft)
1	30	13	3	4	13	0	0	0	
2	38	13	4	6	13	0	0	0	
3	38	20	8	13	19	5	13	0	
4 except marine	49	20	8	13	19	10	13	10	2
5 and marine 4	49	20	13	17	30	15	19	10	2
6	49	20+5[a]	15	20	30	15	19	10	4
7 and 8	49	20+5[a]	19	21	38	15	19	10	4

[Ref. Table N1102.1.2]

a: R-20 cavity insulation plus R-5 continuous insulation

TABLE 15-2 Alternative insulation *R*-value requirements by component

Climate zone	Ceiling w/attic space *R*-value		Ceiling w/o attic space *R*-value		Wood-frame wall w/both cavity and continuous insulation *R*-value		Floor *R*-value	
	Standard truss or rafters	Energy-type truss	General rule	Max. 500 sf or 20% ceiling area	Cavity insulation	Continuous insulation sheathing	General rule	Insulation filling cavity
1	30	30	30	–	–	–	13	13
2	38	30	38	30	–	–	13	13
3	38	30	38	30	13	5	19	19
4 except marine	49	38	49	30	13	5	19	19
5 and marine 4	49	38	49	30	13	5	30	19
6	49	38	49	30	13	10	30	19
7 and 8	49	38	49	30	13	10	38	19

a: R-20 cavity insulation plus R-5 continuous insulation

Slab-on-grade floors

At the perimeter of the thermal envelope, slabs with a floor surface less than 12 inches below grade require insulation with the specified R-value and minimum depth shown in Table 15-1. Heated slabs require an additional R-5 of insulation under the entire slab. Perimeter rigid insulation placed horizontally beneath a slab performs the same function as vertical insulation placed alongside the foundation. Therefore, the code permits the depth requirement to be satisfied by a combination of vertical and horizontal insulation. Because termites will readily excavate through foam plastic to create a path to wood construction, geographic locations subject to very heavy termite infestation are exempt from the slab-edge insulation requirements (Figure 15-5). [Ref. N1102.2.10]

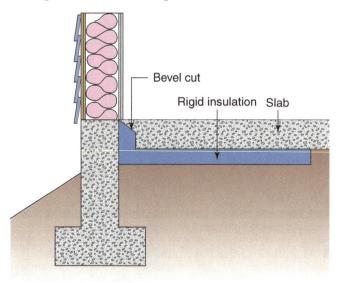

FIGURE 15-5 Slab-on-grade insulation

Crawl space walls

The IRC provides two options for the insulation of crawl spaces (referred to as underfloor spaces in other parts of the code): insulation of the floor above the crawl space or, when the crawl space is not ventilated to the outside, insulation of the exterior walls. For the latter, the IRC provides the minimum R-value for either cavity or continuous wall insulation, as shown in Table 15-1, and requires the insulation to extend vertically to a point 2 feet below the interior grade or horizontally 2 feet. The code also specifies requirements for a vapor retarder on exposed earth of unventilated crawl spaces (Figure 15-6). See Chapter 5 of this publication for underfloor space ventilation and access requirements, and Chapter 9 for foam plastic insulation provisions. [Ref. N1102.2.11]

WINDOWS AND DOORS

Exterior windows, doors and skylights are referred to as fenestration products and must meet the prescribed energy efficiency requirements. The

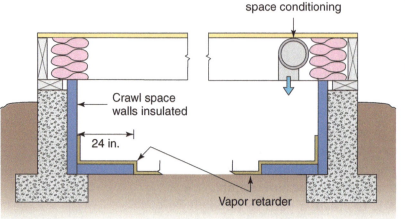

FIGURE 15-6 Insulation for unvented crawl spaces

fenestration requirements apply to both glazed doors and opaque doors. The values of the *U*-factor and solar heat gain coefficient (SHGC) for fenestration products are determined by the referenced standards of the National Fenestration Rating Council (NFRC). For products that lack a label certifying conformance to the applicable NFRC standard, the IRC assigns default values based on the construction of the product. The values in Table 15-3 are maximum *U*-factors and SHGCs for each component. In addition, the IRC regulates air infiltration rates of fenestration products, which are also measured and labeled in accordance with NFRC standards. The maximum infiltration rate for windows, skylights and sliding glass doors is 0.3 cubic feet per minute (cfm) per square foot. For swinging doors, the maximum value is increased to 0.5 cfm per square foot (Figure 15-7). **[Ref. N1102.3, N1102.4.3]**

> ## You Should Know
>
> ### *U*-factor
>
> The *U*-factor is a measurement of heat transmission through building components such as windows and doors. As opposed to *R*-values, a higher *U*-factor indicates a greater transmission of heat and a decreased effectiveness to conserve energy. ●

TABLE 15-3 *Fenestration requirements by component*

Climate zone	Fenestration *U*-factor	Skylight *U*-factor	Glazed fenestration SHGC
1	NR	0.75	0.25
2	0.40	0.65	0.25
3	0.32	0.55	0.25
4 except marine	0.32	0.55	0.40
5 and marine 4	0.30	0.55	NR
6	0.30	0.55	NR
7 and 8	0.30	0.55	NR

[Ref. Table N1102.1.2]

FIGURE 15-7 Window NFRC energy performance label

AIR LEAKAGE

The energy conservation provisions intend to limit air leakage and infiltration through the building thermal envelope by requiring a continuous air barrier and the sealing of any breaks or penetrations in that air barrier. Testing is required to verify compliance with the air leakage requirements.

Sealing

The building thermal envelope relies on properly installed insulation fit snugly to fill all gaps and a continuous air barrier to conserve energy (Figures 15-8 and 15-9). All joints, seams, and penetrations, including utility penetrations, of the building thermal envelope must be sealed against air movement. The code is specific in listing additional locations that are most vulnerable to air leakage and require sealing (Figure 15-10). These include

- Openings around window and door assemblies
- Junctions in wall framing
- Garage separation from conditioned spaces
- Behind tubs and showers on exterior walls
- Attic access openings
- Rim joists
- Recessed lighting
- Shafts and penetrations to the exterior or unconditioned space
- Plumbing and wiring
- Electrical boxes on exterior walls
- HVAC register boots
- Concealed sprinklers

Wood-burning fireplaces are also identified as significant sources of air leakage, and the code requires tight-fitting flue dampers or doors and outdoor combustion air. If doors are used, they must be listed and labeled for the fireplace. [Ref. N1102.4]

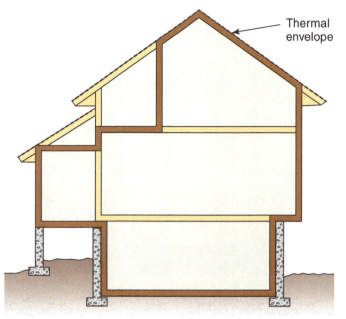

Thermal envelope

FIGURE 15-8 The building thermal envelope is an assembly of elements that provide a boundary between conditioned space and unconditioned space.

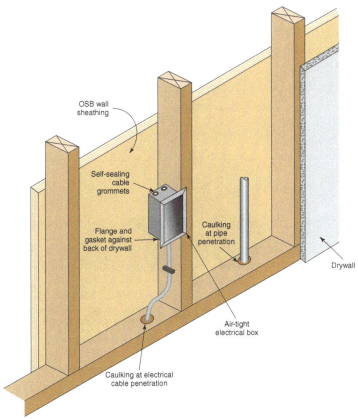

FIGURE 15-9 Components of a continuous air barrier

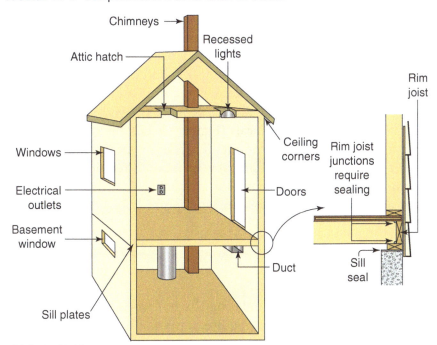

FIGURE 15-10 Typical sources of air leakage

Recessed luminaires

Recessed luminaires (light fixtures) can be a major source of heat loss and moisture introduced into attic spaces. Often referred to as "can lights," recessed luminaires installed in the thermal envelope, typically an insulated ceiling, require a tight seal to limit air leakage. The IRC requires an IC (insulation contact) fixture tested and labeled to conform to the specified air movement standard (Figure 15-11). **[Ref. N1102.4.5]**

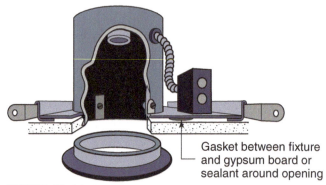

Type IC rated fixture
labeled as meeting
ASTM E 283 with 2.0 CFM
maximum air movement

Gasket between fixture
and gypsum board or
sealant around opening

FIGURE 15-11 Recessed luminaires in thermal envelope

Testing

The code requires blower door testing of all dwelling units to determine compliance with the established limits on air leakage through the building thermal envelope. The maximum air leakage rate varies according to climate zone. Testing in the warmest geographic locations in Climate Zones 1 and 2 must demonstrate that air leakage does not exceed 5 air changes per hour. In colder climates where comfort heating is used to varying degrees in winter, Zones 3 through 8, a tighter building thermal envelope is required and the threshold is set at 3 air changes per hour. Blower door testing is conducted at the prescribed pressure of 50 Pascals (Figure 15-12). Because such tight construction reduces the amount of fresh air infiltrating into the residence, whole-house mechanical ventilation is required to bring in outdoor air and improve the indoor air quality. See Chapters 10 and 12 for additional information on ventilation. **[Ref. N1102.4.1.2, N1103.6]**

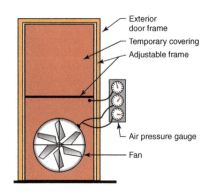

Exterior
door frame

Temporary covering

Adjustable frame

Air pressure gauge

Fan

FIGURE 15-12 The air leakage rate of the building thermal envelope is tested with a blower door.

SYSTEMS

In addition to provisions regulating the building thermal envelope, the code also sets minimum requirements for certain components of mechanical, plumbing, and lighting systems to conserve energy. For example, a programmable thermostat is required for the primary heating and cooling system. Occupants then have the ability to conserve energy through optimum temperature settings on a daily schedule. **[Ref. N1103.1]**

Duct insulation and sealing

Supply and return ducts installed in attics require at least R-8 insulation. This may be reduced to R-6 for ducts in other locations. Ducts entirely within conditioned spaces do not require insulation. The code requires sealing of all ducts in accordance with the mechanical provisions of the IRC. Unless ducts and air handlers are located entirely within the building thermal envelope, air testing is required to verify compliance with the total leakage criteria under the prescribed test pressures. Because building framing cavities are difficult to seal and impair the efficiency of the HVAC system, stud cavities and enclosed floor joist spaces are not permitted to serve as ducts or air plenums for supply or return air. [Ref. N1103.3]

Piping insulation

Piping insulation with a minimum value of R-3 is required for mechanical system piping (such as hydronic heating or cooling tubing) that is designed to carry fluids above 105°F or below 55°F. Similarly, hot water piping in many cases must also be insulated to a minimum of R-3. This insulation is always required for piping exceeding ¾-inch in diameter and for piping in the following locations:
- From the water heater to a distribution manifold
- Underground or under slab
- Outside the conditioned space.
   [Ref. N1103.4, N1103.5]

Lighting

High-efficacy lamps (typically LED or compact fluorescent lamps) in a residential setting save electrical energy when compared to incandescent lighting. The code requires high-efficacy lamps in at least 90 percent of permanent lighting fixtures in dwelling units. [Ref. N1104]

ENERGY CERTIFICATE

The IRC requires the builder or registered design professional to complete an energy efficiency certificate, listing the installed insulation and fenestration values. The certificate must also list the type and efficiency of installed heating, cooling and water heating equipment. Because electric furnaces, baseboard heaters and unvented gas-fired heaters do not provide the lowest energy consumption when compared to other methods of comfort heating and their energy efficiency ratings may be misleading, the IRC requires such appliances to be individually listed on the certificate without an efficiency designation. In addition, results of the blower door test and duct testing must appear on the certificate. The code stipulates posting of the certificate in a furnace room, a utility room or another approved location inside the building. If approved, the permanent certificate can be affixed to an interior electrical service panel, but cannot cover the service directory or other required information governed by the electrical code (Figure 15-13). [Ref. N1101.14]

Energy Efficiency Certificate

Insulation rating	R-Value	
Ceiling/Roof	59.00	
Wall	30.00	
Floor/Foundation	15.00	
Ductwork (unconditioned spaces):	_____	

Glass & door rating	U-Factor	SHGC
Window	0.28	0.40
Door	0.35	NA

Heating & cooling equipment	Efficiency	
Heating system: _____	_____	
Cooling system: _____	_____	
Water heater: _____	_____	

Building air leakage and duct test results	
Building air leakage test results	_____
Name of air leakage tester	_____
Duct tightness test results	_____
Name of duct tester	_____

Name: _____ Date: _____

Comments: _____
_____

FIGURE 15-13 Permanent energy certificate

Protection from Other Hazards

Chapter 16: Other Hazards of the Built Environment

Other Hazards of the Built Environment

This chapter provides information on structural and environmental hazards often associated with dwelling and accessory building construction regulated by the *International Residential Code* (IRC). The IRC identifies regions of the country most vulnerable to termite infestation and prescribes appropriate measures to protect against termite damage in those areas. IRC Section R326 references the *International Swimming Pool and Spa Code* (ISPSC) for design and construction of residential swimming pools and hot tubs including provisions for protective barriers. Requirements for radon control appear in IRC Appendix F and are not in force unless specifically adopted by the jurisdiction. Airborne lead and asbestos hazards are encountered primarily in building remodeling or rehabilitation and are usually governed by state regulations. The IRC and federal regulations limit lead content in water piping and fittings for new construction.

TERMITE PROTECTION

Subterranean termites cause significant damage to concealed structural and nonstructural wood components. They thrive in moist ground and usually invade homes by building mud tunnels on the surface of foundations. They may also travel inside hollow block masonry, through plastic foam insulation or directly through untreated wood in contact with the ground or organic materials such as mulch. Mud tunnels may be observed only when the minimum clearances above grade are maintained as required by the IRC. Termite activity may also be deterred by blocking their access points above the foundation with termite shields and pressure-preservative-treated foundation plates.

The termite protection provisions of the IRC are applicable to all geographic areas subject to termite damage. When adopting the IRC, the jurisdiction designates the level of hazard based on a history of local subterranean termite damage. The infestation probability map in Figure 16-1 shows the approximate areas of hazard rated from none or slight to very heavy. **[Ref. R301.2, Table R301.2(1), Figure R301.2(6)]**

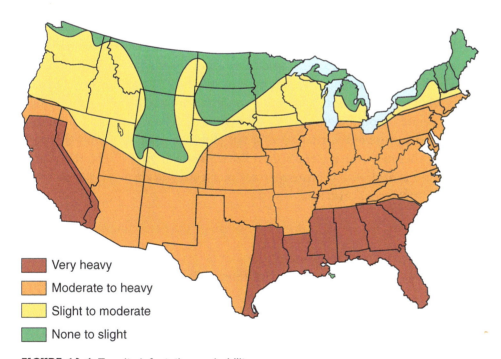

Very heavy
Moderate to heavy
Slight to moderate
None to slight

FIGURE 16-1 Termite infestation probability map

Termite control methods

For wood frame construction, the IRC requires one or more of six methods of protection in all areas subject to termite damage: chemical termiticide treatment, termite baiting, pressure-preservative-treated wood, naturally termite-resistant wood, cold-formed steel framing and physical barriers. Liquid termiticide treatment is the most common termite control method. These chemicals are placed in the ground to kill or repel termites. Formulations are available to treat wood on-site. Baiting systems are used when application of a liquid termiticide is not feasible or practical. The IRC requires pressure-preservative-treated wood or naturally durable wood in the locations described in Chapter 6 of this publication for protection against decay and termites. Physical barriers are typically continuous corrosion-resistant metal termite shields placed at the top of foundation walls or piers. When installed on exterior foundation walls, termite shields have limited effectiveness and require one or more additional approved methods. [Ref. R318.1, R318.3]

Foam plastic

In geographic locations where the termite probability designation is "very heavy," the IRC generally prohibits unprotected foam plastic insulation on the exterior surface of foundation walls or under slab foundations on grade. [Ref. R318.4]

SWIMMING POOLS AND HOT TUBS

The IRC references the *International Swimming Pool and Spa Code* (ISPSC) for regulating the design and construction of swimming pools and spas (hot tubs) for one- and two-family dwellings and townhouses. The ISPSC establishes minimum requirements to provide a reasonable level of safety related to the use of pools and spas, including provisions for related piping, equipment and materials. The barrier requirements intend to prevent unattended small children from entering the pool area, thereby reducing the risk of drowning or injury. Entrapment protection is addressed to prevent serious injury or drowning associated with pump suction. [Ref. R326]

Barriers

The barrier provisions apply to outdoor and indoor swimming pools and spas (hot tubs). An approved lockable safety cover for a hot tub satisfies the barrier requirement. For swimming pools, a powered safety cover complying with ASTM F 1346 is permitted to serve as the barrier. Otherwise, required pool barriers such as wood or chain-link fences must be at least 48 inches high and constructed to prevent climbing by children. Access gates must be self-closing and self-latching with the hardware arranged so the gate cannot be opened from the side opposite the pool. For above-ground pools, the wall of the pool may serve as the barrier, provided it meets the 48-inch height requirement and has a ladder or steps that can be removed or locked to prevent access (Figures 16-2 and 16-3). [Ref. ISPSC 305]

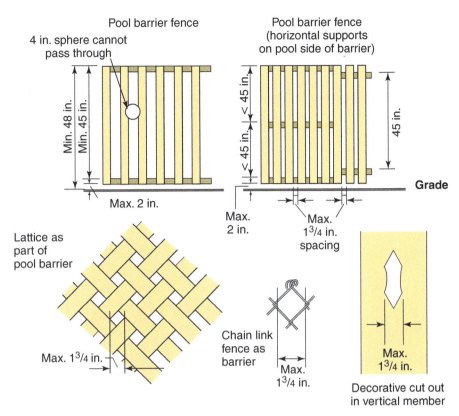

FIGURE 16-2 Swimming pool barrier requirements

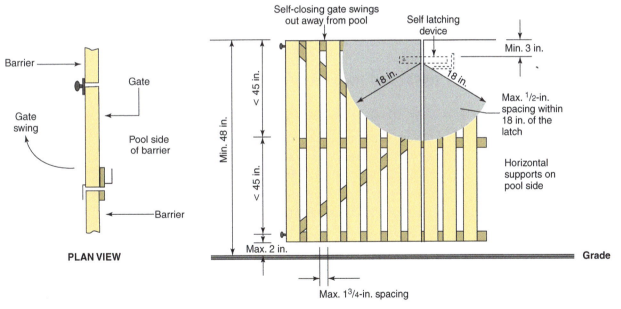

View of a gate from outside

FIGURE 16-3 Pool access gate requirements

In addition to outside walls or fences as barriers, the ISPSC requires protection for occupants of the residence as well. Unless the pool is equipped with an approved powered safety cover, doors and windows with direct access to an indoor or outdoor swimming pool require an audible warning alarm system. The alarm must be listed and labeled as a water hazard entrance alarm in accordance with

UL 2017. The ISPSC also authorizes the building official to accept alternate and equivalent self-closing and self-latching devices in lieu of the door alarm (Figure 16-4). **[Ref. ISPSC 305.4]**

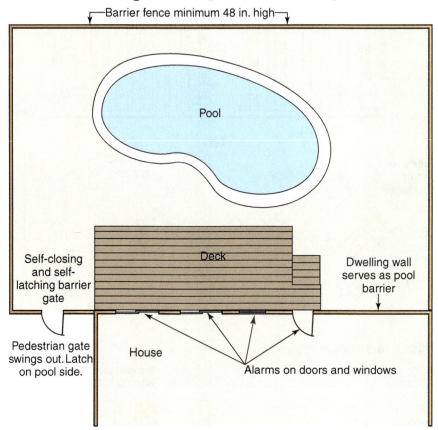

FIGURE 16-4 Pool barrier and door alarm requirements

Entrapment protection

Underwater entrapment occurs when parts of the body cover the suction outlet, creating a vacuum, or when long hair or clothing becomes entangled in a drain without a safety cover, potentially resulting in severe injury or drowning. The ISPSC references ANSI/APSP/ICC-7, American National Standard for Suction Entrapment Avoidance in Pools,

for avoidance of entrapment hazards to improve pool safety. Based on the latest information and technology to address all forms of entrapment hazards and their underlying causes, the code and the referenced standard provide that all swimming pools and spas be equipped with proper anti-entrapment drain covers and circulation and drainage systems. **[Ref. ISPSC 310]**

RADON CONTROL

Appendix Chapter F establishes prescriptive provisions for radon resistant new construction. The prescribed measures act to reduce introduction of soil gases potentially containing radon into the living space of a dwelling and to provide a cost-effective means of mitigation should testing reveal unacceptable levels of radon after construction. A decay product of uranium, radon is a radioactive gas that occurs naturally in the soil in varying concentrations and over time can cause damage to the lungs and increases the potential for developing lung cancer. It is not possible to predict with accuracy or certainty the levels of radon that will occur in a given house until the building is enclosed and tested. Houses in the same neighborhood, because of varying radon concentrations in the soil, may experience very different levels of radon.

The primary means for reducing entry of radon gas into the dwelling is through sub-slab depressurization. This method seals all entry points of the slab and basement foundation wall and vents the area below the slab out through the roof. Effectiveness of this system depends on an air-permeable granular base, such as sand or gravel, below the slab. This passive vent pipe may terminate in a tee fitting below the floor or may connect directly to a gas-tight sump pit or foundation drain tile. Even a passive system without a fan on the vent pipe creates a chimney effect and intends to maintain a lower air pressure below the slab than occurs in the dwelling. The provisions of Appendix F allow for easy conversion to an active system if necessary. If testing in the lowest living space of the dwelling reveals radon levels above the action level, an approved in-line fan may be installed in the vent pipe in the attic. The fan runs continuously to draw air from below the slab. With lower sub-slab pressure, soil gases will follow the path of least resistance out through the roof rather than migrating into the living space (Figures 16-5 through 16-7). **[Ref. AF103]**

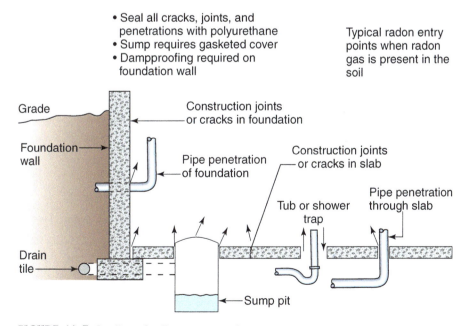

FIGURE 16-5 Sealing of soil gas-entry points

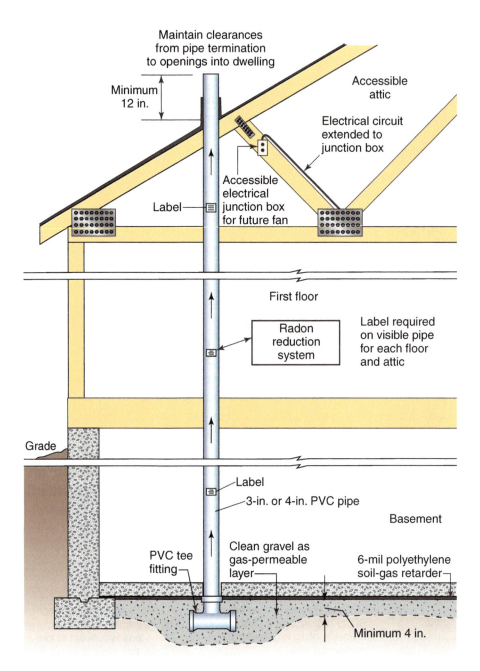

FIGURE 16-6 Passive sub-slab depressurization system

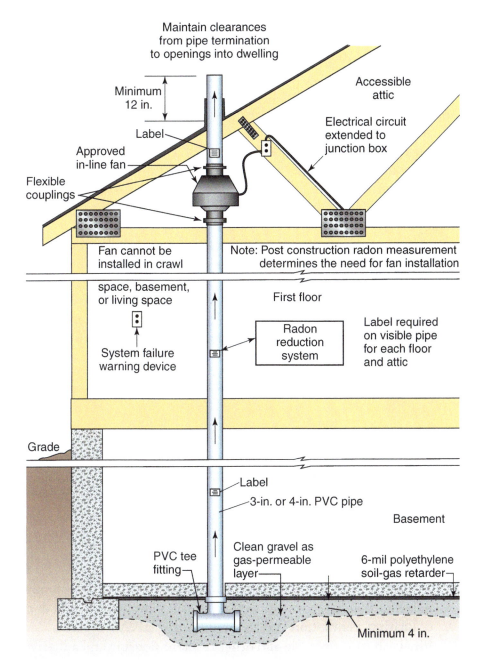

Maintain clearances
from pipe termination
to openings into dwelling

Minimum
12 in.

Accessible
attic

Label

Electrical circuit
extended to
junction box

Approved
in-line fan

Flexible
couplings

Fan cannot be
installed in crawl
space, basement,
or living space

Note: Post construction radon measurement
determines the need for fan installation

First floor

System failure
warning device

Radon
reduction
system

Label required
on visible pipe
for each floor
and attic

Grade

Label

3-in. or 4-in. PVC pipe

Basement

PVC tee
fitting

Clean gravel as
gas-permeable
layer

6-mil polyethylene
soil-gas retarder

Minimum 4 in.

FIGURE 16-7 Active sub-slab depressurization system

LEAD

Lead is a toxic substance that can cause harmful health effects when ingested. Children are particularly susceptible to the effects of lead poisoning, which is most commonly caused by the ingestion of lead paint dust or chips in older houses. A secondary cause is from drinking water containing lead that has leached from lead pipes or solder.

Paint used in homes prior to 1960 typically contained relatively high levels of lead. These levels decreased until 1978, when the Consumer Product Safety Commission banned lead-based paint. Therefore, the risk of lead poisoning is greatest in houses constructed before 1960. Even so, paint surfaces maintained in good condition pose little risk. It is through sanding, scraping, peeling, flaking or chipping that lead paint becomes hazardous. Care must be taken in remodeling, renovation and maintenance to contain and properly dispose of dust and debris.

Communities often implement nuisance or housing ordinances to regulate the maintenance of buildings. The *International Property Maintenance Code* (IPMC) requires all interior and exterior surfaces, including windows, doors and trim, to be maintained in good condition. The IPMC requires repairing, removing or covering any peeling, chipping, flaking or abraded paint. **[Ref. IPMC 304.2, IPMC 305.3]**

Lead is also a concern in drinking water supplied through older piping systems of lead pipe, fittings or solder. Such materials have very limited use in new construction. The IRC regulates lead content in systems of potable-water supply. Solders and fluxes are limited to a maximum of 0.2 percent lead. The code limits lead content to 0.25 percent in pipe and fittings that convey water used for drinking and cooking. **[Ref. P2906.2, P2906.14]**

ASBESTOS

Asbestos fibers are a health hazard when inhaled into the lungs. Modern construction materials no longer contain asbestos, and its use is generally prohibited, even in the repair of existing buildings. For example, while IRC Appendix Chapter J, "Existing Buildings and Structures," and the *International Existing Building Code* (IEBC) generally permit repairs using similar materials, they specifically prohibit the use of asbestos and other hazardous materials. **[Ref. IRC AJ301.1, IEBC 602.2]**

Asbestos is often found in the following materials in older homes:
- Asbestos blanket, paper or tape on steam pipes, boilers and ducts of heating, ventilation and air conditioning (HVAC) systems
- Vinyl asbestos floor tile and other resilient flooring and adhesives
- Cement board sheets used adjacent to wood-burning stoves for protection of combustible materials
- Gypsum board, plaster and ceiling tile
- Patching and joint compounds and spray-on texture materials
- Asbestos cement sheet siding, siding and roofing shingles, and other roofing materials.

The asbestos materials listed above are not hazardous while they remain in place and intact. Drilling, sawing, sanding or removing such products is likely to create hazardous dust and airborne asbestos fibers. Special precautions must be followed to contain and dispose of asbestos materials properly and safely during abatement, repair or renovation. Many states have laws regulating the removal and handling of asbestos-containing materials, as well as related permitting and licensing for such operations.

GLOSSARY

A

accepted engineering practice – Engineered analysis based on well-established principles of mechanics and conforming to accepted principles, tests, or standards of nationally recognized technical authorities.

access (to) – that which enables a device, an appliance or equipment to be reached by ready access or by a means that first requires the removal or movement of a panel, door or similar obstruction

accessory structure – A building or structure accessory to and incidental to the dwelling, such as a detached garage or shed, located on the same lot as the dwelling.

air admittance valve – a one-way valve designed to allow air into the plumbing drainage system where a negative pressure develops in the piping. This device shall close by gravity and seal the terminal under conditions of zero differential pressure (no flow conditions) and under positive internal pressure.

air barrier – one or more materials joined together in a continuous manner to restrict or prevent the passage of air through the building thermal envelope and its assemblies

air gap – Open space between the potable-water outlet and the flood-level rim of the fixture, effectively separating the potable water from the source of contamination.

ampacity – The current, in amperes, that a conductor can carry continuously under the conditions of use without exceeding its temperature rating.

approved – Acceptable to the building official.

approved agency – An established and recognized agency regularly engaged in conducting tests or furnishing inspection services, when such agency has been approved by the building official.

arc-fault circuit interrupter (AFCI) – A device intended to provide protection from the effects of arc faults by recognizing characteristics unique to arcing and by functioning to de-energize the circuit when an arc fault is detected.

AWG – American Wire Gauge; indicates the size of the wire. As the AWG number decreases, the wire size (diameter) increases.

B

bonding – Permanent joining of metallic parts to form an electrically conductive path that will ensure electrical continuity and the capacity to conduct safely any current likely to be imposed.

bonding jumper – A reliable conductor to ensure the required electrical conductivity between metal parts required to be electrically connected.

building drain – Lowest drainage piping inside the house; extends 30 inches beyond the exterior walls to connect to the building sewer. The building drain collects the discharge from all other drainage piping in the dwelling.

building official – The officer or other designated authority charged with the administration and enforcement of the IRC.

building sewer – That part of the drainage system that extends from the end of the building drain 30 inches outside of the building and conveys its discharge to a public sewer, private sewer, individual sewage-disposal system, or other approved point of disposal.

building thermal envelope – The basement walls, exterior walls, floors, ceilings, roofs and any other building element assemblies that enclose conditioned space or provide a boundary between conditioned space and exempt or unconditioned space.

C

Category IV vented appliance – An appliance that operates with a positive vent static pressure and with a vent gas temperature that is capable of causing excessive condensate production in the vent. Category IV appliances are referred to as very high efficiency condensing appliances. They rely on mechanical means rather than gravity to vent the low-temperature flue gases to the outside. Vents must be of noncorrosive material as specified by the appliance manufacturer, such as PVC pipe.

crawl space – An underfloor space that is not a basement.

cricket – A sloped flashing on the up-roof side of a chimney to divert water from above the chimney to each side.

D

dead loads – The actual weight of all materials of construction incorporated into the building, including but not limited to walls, floors, roofs, ceilings, stairways, built-in partitions, finishes, cladding and fixed service equipment. Dead loads are fixed and considered permanent.

design flood elevation – The elevation subject to a 1 percent or greater chance of flooding in any year or the otherwise legally designated elevation in a flood hazard area.

d.f.u. – In plumbing, drainage fixture units. A measure of probable discharge into the drainage system by various types of plumbing fixtures, used to size DWV piping systems. The drainage fixture unit value for a particular fixture depends on its volume rate of drainage discharge, on the time duration of a single drainage operation, and on the average time between successive operations.

direct-vent appliance – Fuel-burning appliance with a sealed combustion system that draws all air for combustion from the outside atmosphere and discharges all flue gases to the outside atmosphere.

draft stop – Material or construction installed to restrict the movement of air within open spaces of concealed areas of buildings.

dwelling – A building that contains one or two dwelling units occupied or intended to be occupied for living purposes.

dwelling unit – A single unit providing complete independent living facilities for one or more persons, including permanent provisions for living, sleeping, eating, cooking and sanitation.

DWV – The drainage, waste and vent (DWV) system consists of all piping for conveying wastes from plumbing fixtures, appliances and appurtenances, including fixture traps and venting systems.

E

effectively grounded – An electrical term meaning intentionally connected to earth through a ground connection or connections of sufficiently low impedance and having sufficient current-carrying capacity to prevent the buildup of voltages that may result in undue hazards to connected equipment or to persons.

emergency escape and rescue opening – An operable exterior window, door or similar device that provides for a means of escape and access for rescue in the event of an emergency.

equipment grounding conductor – Conductor used to connect the non-current-carrying metal parts of equipment, conduit, and other enclosures to the system grounded (neutral) conductor, the grounding electrode conductor or both, at the service equipment.

Evaluation Service (ES) report – A report that presents the findings of ICC-ES as to the compliance with code requirements of the subject of the report—a particular building product, component, method or material.

F

fenestration – Skylights, roof windows, vertical windows (whether fixed or moveable), opaque doors, glazed doors, glass block and combination opaque/glazed doors.

fireblocking – Building materials installed to resist the free passage of flame to other areas of the building through concealed spaces.

fire separation distance – The distance measured from the building face to the closest interior lot line; to the centerline of a street, an alley or public way; or to an imaginary line between two buildings on the lot. The distance is measured at a right angle from the face of the wall.

flood hazard area – The area within a flood plain subject to a 1 percent or greater chance of flooding in any year, or the area otherwise legally designated as a flood hazard area.

floodway – The channel of the river, creek, or other watercourse and the adjacent land areas that must be reserved in order to discharge the base flood without cumulatively increasing the water surface elevation more than a designated height.

G

GFCI – Ground-fault circuit interrupter. A device intended to protect people, it functions to de-energize a circuit within an established period of time when a current to ground exceeds a certain value, indicating a fault.

gpm – Gallons per minute.

grade floor opening – A window or other opening located such that the sill height of the opening is not more than 44 inches above or below the finished ground level adjacent to the opening.

grade plane – A reference plane representing the average of the finished ground level adjoining the building at all exterior walls. Where the finished ground level slopes away from the exterior walls, the reference plane shall be established by the lowest points within the area between the building and the lot line or, where the lot line is more than 6 feet from the building between the structure and a point 6 feet from the building.

gray water – Waste discharged from lavatories, bathtubs, showers, clothes washers and laundry trays.

ground – A conducting connection between an electrical circuit or equipment and the earth.

grounded conductor – System or circuit conductor that is intentionally grounded. This is commonly referred to as the *neutral* and is sometimes confused with the terminology for grounding conductor. Grounded conductors (wires) generally are marked with white.

grounding conductor – A conductor used to connect equipment or the grounded circuit of a wiring system to a grounding electrode or electrodes. The grounding conductor typically bonds to the grounded (neutral) conductor only at the service equipment, and typically has green markings or is bare wire.

grounding electrode – Device that establishes an electrical connection to earth. Ground rods, concrete-encased reinforcing bars and underground copper water service piping are examples of grounding electrodes.

grounding electrode conductor – Conductor used to connect the grounding electrode(s) to the equipment grounding conductor, to the grounded conductor or to both, typically at the service.

gypsum board – The generic name for a family of sheet products consisting of a noncombustible core primarily of gypsum with paper surfacing. Gypsum wallboard, gypsum sheathing, gypsum base for gypsum veneer plaster, exterior gypsum soffit board and water-resistant gypsum backing board are types of gypsum board.

H

habitable attic – A finished or unfinished habitable space within an attic.

habitable space – Space in a building for living, sleeping, eating, or cooking. Bathrooms, toilet rooms, closets, halls, storage or utility spaces, and similar areas are not considered habitable spaces.

high-efficacy lamps – Compact fluorescent lamps, light-emitting diode (LED) lamps, T-8 or smaller-diameter linear fluorescent lamps or lamps with a minimum efficacy of (1) 60 lumens per watt for lamps over 40 watts; (2) 50 lumens per watt for lamps over 15 watts to 40 watts; and (3) 40 lumens per watt for lamps 15 watts or less.

I

inertia – The property of matter that produces a tendency of objects at rest to remain at rest and objects in uniform motion to remain in uniform motion.

interlayment – In the application of wood shakes, interlayment is an 18-inch-wide strip of at least No. 30 felt shingled between each course of shakes in such a manner that no felt is exposed to the weather.

J

jalousie – A window, blind, or shutter having adjustable horizontal slats. A jalousie window has overlapping glass slats which open to allow the passage of air and light.

K

kcmil – Thousands of circular mils. In the North American electrical industry, conductors larger than 4/0 AWG are generally identified by the area in thousands of circular mils (kcmil), where 1 kcmil = 0.5067 mm^2. An older abbreviation for 1,000 circular mils is *mcm*.

L

labeled – Devices, equipment, or materials bearing a label, seal, symbol or other identifying mark of a testing laboratory that attests to compliance with a specific standard.

light-frame construction – A type of construction whose vertical and horizontal structural elements are formed primarily by a system of repetitive wood or cold-formed steel framing members.

listed, listing – Terms referring to equipment that is listed by an approved testing agency as complying with nationally recognized standards when installed in accordance with the manufacturer's installation instructions.

live loads – Loads produced by the use and occupancy of the building, such as people and furnishings, and not including construction or environmental loads such as those for wind, snow, rain, earthquake, flood or dead loads. Live loads are variable and temporary.

loads – Forces that are applied to the structural system of the building.

local exhaust – An exhaust system that uses one or more fans to exhaust air from a specific room or rooms within a dwelling.

luminaire – Complete lighting unit (lighting fixture), consisting of a lamp or lamps, together with parts designed to distribute the light, to position and protect the lamps and ballast, and to connect the lamps to the power supply.

M

manufactured home – A structure, transportable in one or more sections, which in the traveling mode is 8 body feet or more in width or 40 body feet or more in length, or, when erected on site, is 320 square feet or more, and which is built on a permanent chassis and designed to be used as a dwelling with or without a permanent foundation. A mobile home is considered a manufactured home.

means of egress – The path of travel from any occupied portion of the building to the outdoors. Stairways, ramps, landings, hallways and doors are components of the means of egress.

multipurpose fire sprinkler system – A system that supplies domestic water to both plumbing fixtures and fire sprinklers.

O

on-site nonpotable water reuse systems – Water systems for the collection, treatment, storage, distribution and reuse of nonpotable water generated on site, including but not limited to graywater systems. This definition does not include rainwater harvesting systems.

overcurrent – Any current in excess of the rated current of equipment or the ampacity of a conductor. It may result from overload, short circuit or ground fault.

overcurrent protection device – A circuit breaker, fuse or other device that protects the circuit by opening the device, thereby disconnecting power to the circuit, when the current reaches a value that will cause an excessive or dangerous temperature rise in conductors (overcurrent).

P

pan flashing – Corrosion-resistant flashing at the base of an opening that is integrated into the building exterior wall to direct water to the exterior.

R

ready access (to) – That which enables a device, appliance or equipment to be directly reached without requiring the removal or movement of any panel, door or similar obstruction.

registered design professional – An individual, such as an engineer or architect, who is registered or licensed to practice a design profession as defined by the statutory requirements of the professional registration laws of the state or jurisdiction in which the project is to be constructed.

S

seismic – Characteristic of or having to do with earthquake ground motion.

stack vent – Continuation of the stack above the highest horizontal drain connection.

substantial improvement – Any repair, reconstruction, rehabilitation, addition, or improvement of a building or structure, the cost of which equals or exceeds 50 percent of the market value of the structure before the improvement or repair is started. A determination of substantial improvement relates to buildings located in a flood hazard area.

T

townhouse – A single-family dwelling unit constructed in a group of three or more attached units in which each unit extends from foundation to roof and with a yard or public way on at least two sides.

U

ungrounded conductor – Often referred to as the "hot" conductor (wire). The insulation color is typically black or red but may be of any color other than white, green or gray.

W

waste stack – A main line of vertical DWV piping that conveys only liquid sewage not containing fecal material.

whole-house mechanical ventilation system – An exhaust system, supply system, or combination thereof that is designed to mechanically exchange indoor air for outdoor air when operating continuously or through a programmed intermittent schedule to satisfy the whole-house ventilation rate.

wood structural panel – A panel manufactured from veneers or wood strands or wafers bonded together with waterproof synthetic resins or other suitable bonding systems. Examples of wood structural panels are plywood, OSB and composite panels.

INDEX

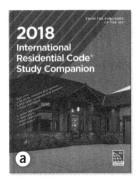   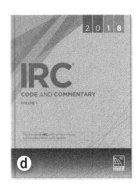 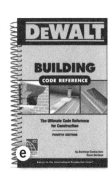

Helpful tools for Your Residential Code

a. 2018 International Residential Code® Study Companion

The Study Companion is organized into study sessions that include learning objectives, key points for review, code text, hundreds of illustrations and applicable code commentary. The sessions discuss the building, mechanical, plumbing, fuel gas and electrical provisions of the 2018 IRC®. Particular emphasis is placed on the building planning requirements of Chapter 3 and the floor, wall, ceiling, and roof framing provisions.

Features:

• 18 study sessions

• 630 total study questions with answer key

SOFT COVER   #4117S18
PDF DOWNLOAD   #8950P838

b. Residential Code Essentials: Based on the 2018 IRC

Detailed full-color illustrations enhance comprehension of code provisions

Explores those code provisions essential to understanding the application of the IRC in a straightforward and easy-to-read format. The user-friendly approach of the text simplifies critical concepts so that users can achieve a more complete understanding of the code's intent. Full-color illustrations, examples and simplified tables assist the reader in visualizing the code requirements. This up-to-date, step-by-step guide's topic organization reflects the intent of the IRC and facilitates understanding and application of the code provisions.

SOFT COVER   #4131S18
PDF DOWNLOAD   #8951P013

FROM APA & ICC!

c. A Guide to the 2018 IRC Wood Wall Bracing Provisions

Although it is of critical importance when designing, performing plan review, building or inspecting a structure, wall bracing is a common source of confusion and misapplication. This illustrative guide was developed to help building designers, builders, building officials and others using the code in the application of the lateral bracing requirements of the 2018 IRC.

SOFT COVER   #7102S18
PDF DOWNLOAD   #8799P18

d. 2018 IRC® Code and Commentary, Volumes 1 & 2 (Chapters 1–44)

This helpful set contains the full text of 2018 IRC, including tables and figures, followed by corresponding commentary at the end of each section to help code users understand the intent of the code provisions and learn how to apply them effectively. Volumes also sold separately.

SOFT COVER   #3110S18
PDF DOWNLOAD   #871P18
SOFT + PDF COMBO   #3110SP18

e. DeWALT Building Code Reference, Fourth Edition

A simple, easy-to-understand approach to the 2018 IRC that provides illustrations and clear, concise text. Coverage ranges from wall, floor, and roof framing to foundations and footings, containing all the information you need to be successful in the industry in a compact, easy-to-use reference guide. Packaged in a conveniently-sized, durable format, it will withstand a variety of on-the-job trainings and ultimately the wear and tear of jobsites.

SOFT COVER   #9511S18

20-18301

Proctored Remote Online Testing Option (PRONTO™)

Convenient, Reliable, and Secure Certification Exams

Take the Test at Your Location

Take advantage of ICC PRONTO, an industry leading, secure online exam delivery service. PRONTO allows you to take ICC Certification exams at your convenience in the privacy of your own home, office or other secure location. Plus, you'll know your pass/fail status immediately upon completion.

 With PRONTO, ICC's Proctored Remote Online Testing Option, take your ICC Certification exam from any location with high-speed internet access.

 With online proctoring and exam security features you can be confident in the integrity of the testing process and exam results.

 Plan your exam for the day and time most convenient for you. PRONTO is available 24/7.

 Eliminate the waiting period and get your results in private immediately upon exam completion.

 ICC was the first model code organization to offer secured online proctored exams—part of our commitment to offering the latest technology-based solutions to help building and code professionals succeed and advance. We continue to expand our catalog of PRONTO exam offerings.

Discover ICC PRONTO and the wealth of certification opportunities available to advance your career: www.iccsafe.org/MeetPRONTO

ISO/IEC 17065
Product Certification Body
#1000

20-18842

Training and Education

The Learning Center at ICC provides training and education to building safety, fire, design, and construction professionals on how best to apply and enforce the codes and leadership topics.

Training and education programs are designed for maximum impact and results by assisting you in developing professional skills, advancing your career, expanding your knowledge base, and/or preparing for your next certification exam. Stay on the leading edge of the industry while earning valuable CEUs.

All Learning Center training opportunities are developed and led by nationally recognized code experts with decades of experience. We offer convenient learning opportunities in a variety of formats including classroom, virtual, live web session and online courses.

CLASSROOM

The Learning Center's multi-day Institute, single-day Seminars, and Certification Test Academies offer in-depth training from leading code and building safety experts held at locations throughout the U.S. Network with your peers and get on-the-job questions answered at an in person training. Also, Hire ICC to Teach provides the opportunity to bring ICC's expert instructors to the location of your choice. Choose from a wide range of topics or work with the Learning Center experts to develop customized training to meet your needs.

VIRTUAL CLASSROOM

Join a live training event from any location – whether you're in the office or at home. All you need is a standard internet connection. Virtual classrooms are a hybrid learning environment where you participate remotely and experience the same collaboration, instructor interaction and learning benefits as if you were physically in the classroom.

LIVE WEB SESSION

Participate in targeted training events by taking advantage of a web session series to give you the training you need in the timeframe you need it while getting all the CEUs you need for certification renewal.

ONLINE

Online learning is available 24/7 and features topics such as exam study, code training, management, leadership, and financial planning.

PREFERRED PROVIDER PROGRAM

Open the door to extensive training opportunities with ICC-approved educational offerings by a variety of providers as they relate to codes, standards and guidelines, as well as building construction materials, products and methods.

For information about the Learning Center: visit **learn.iccsafe.org** or call **888-422-7233, x33821**